Environmental Science and Engineering

The ultimate goal of this series is to contribute to the protection of our environment, which calls for both profound research and the ongoing development of solutions and measurements by experts in the field. Accordingly, the series promotes not only a deeper understanding of environmental processes and the evaluation of management strategies, but also design and technology aimed at improving environmental quality. Books focusing on the former are published in the subseries Environmental Science, those focusing on the latter in the subseries Environmental Engineering.

Jan Baeyens · Raf Dewil · Barbara Rossi · Yimin Deng
Editors

Proceedings of 2022 4th International Conference on Environment Sciences and Renewable Energy

Selected Topics on New Developments in Environmental and Energy Technologies

Springer

Editors
Jan Baeyens
Beijing Advanced Innovation Centre for Smart Matter Science and Engineering
Beijing University of Chemical Technology
Beijing, China

Barbara Rossi
Department of Engineering Science
University of Oxford
Oxford, UK

Raf Dewil
Department of Engineering Science
University of Oxford
Oxford, UK

Department of Chemical Engineering
KU Leuven
Leuven, Belgium

Yimin Deng
Department of Chemical Engineering
KU Leuven
Leuven, Belgium

ISSN 1863-5520 ISSN 1863-5539 (electronic)
Environmental Science and Engineering
ISBN 978-981-19-9439-5 ISBN 978-981-19-9440-1 (eBook)
https://doi.org/10.1007/978-981-19-9440-1

This Springer imprint is published by the registered company Springer Nature Singapore Pte Ltd.
The registered company address is: 152 Beach Road, #21-01/04 Gateway East, Singapore 189721, Singapore

Conference Committees

Conference Co-Chairs

Jan Baeyens, Beijing University of Chemical Technology, China; KU Leuven, Belgium
Franz Winter, Vienna University of Technology, Vienna, Austria.

Conference Program Co-Chairs

Hongwei Wu, University of Hertfordshire, United Kingdom
Farhad Shahnia, Murdoch University, Australia.

Technical Committees

Catalina Iticescu, University of Galati, Romania
Md. Hasanuzzaman, University of Malaya, Malaysia
Cao Yanpeng, Zhejiang University, China
Hongwei Wu, University of Hertfordshire, United Kingdom
Agus Sofyan, Big Sandy Community and Technical College, USA
Mohamed Ahmed Ibrahim Ahmed, Assiut University, Egypt
Farhad Shahnia, Murdoch University, Australia
Lee Hwang Sheng, University Tunku Abdul Rahman, Malaysia
Rajeev Pratap Singh, Banaras Hindu University, Malaysia
Zhang Yaning, Harbin Institute of Technology, China
Chow Ming Fai, Monash University Malaysia, Malaysia
Chunhua Liu, City University of Hong Kong, China
Ramesh A/L T Subramaniam, Universiti Malaya, Malaysia

Preface I

The 2022 4th International Conference on Environmental Sciences and Renewable Energy (ESRE 2022) has been successfully held in virtual mode this year, whose focus domains were threefold: (1) Environmental Science and Ecological Protection, (2) Environmental Science and Ecological Protection, and (3) Renewable Energy Technology and Energy Chemical Engineering. Among more than 50 papers presented at the conference, the most important ones were subjected to peer-review and selected for inclusion in the Springer Book Series. The selected papers provide new insights into various topics of the target focus domains. They are hence divided into the three subtopics.

(1) **Environmental Science and Ecological Protection** present some novel and potential approaches. With humanity being responsible for many changes that take place on our planet, the study of the environment and the planet, through environmental science and ecology, will foster us to become better stewards of Earth.

In general, both environmental science and ecology play a crucial role in the important protection and preservation of our planet and its organisms for future generations.

While environmental science aims at protecting human beings and the environment from unacceptable environmental factors such as Greenhouse Gas emissions and overall pollution, ecological protection specifically studies how living organisms interact. Ecology, as a special branch of environmental science, provides insight into how ecosystems develop, how humans negatively impact those ecosystems, and how that impact can be minimized. Selected papers, as listed in Part 1 of Table of Contents, deal with specific topics in these fields.

(2) **Environmental Science and Ecological Protection** remain major research topics. The Sustainable Development Goals (SDGs) set the standards, the challenges and the cross-cutting issues that should be met on a worldwide basis. During the conference, different papers were presented to promote the achievement of the SDGs through more profound waste management and waste utilization. These papers are grouped into Part 2 of Table of Contents.

(3) **Renewable Energy Technology and Energy Chemical Engineering**, although already well established, have developed some novel insights. Renewable

energy is energy collected from renewable resources that are naturally available on a human timescale. It includes biomass, biofuels, solar energy, wind, rain, tides, waves and geothermal heat. Recently, hydrogen derived from or using "green" resources has attracted much interest. Although solar photovoltaic and on-/offshore wind are the cheapest newly built power generation facilities, other renewable energy technologies are continuously developed and improved. Renewable energy projects are commonly large-scale, but they can also be applied in rural and remote areas and in developing countries, where energy is often crucial in community development.

Selected papers, as listed in Part 3 of Table of Contents, deal with specific topics in these fields and introduce new concepts in, e.g., concentrated solar power and ammonia-derived hydrogen.

Beijing, China — Jan Baeyens
Sint-Katelijne-Waver, Belgium\Oxford, UK — Raf Dewil
Oxford, UK — Barbara Rossi
Sint-Katelijne-Waver, Belgium — Yimin Deng

Preface II

The 2022 4th International Conference on Environment Sciences and Renewable Energy (ESRE 2022) was held as a virtual conference during May 27–29, 2022. The conference became a unique platform for leading scientists, teachers, experts and practitioners active in both important fields. Conference participants were offered a unique opportunity to present the results of their breakthrough research, to exchange knowledge and experiences, and to discuss current research problems and trends.

The international conference was divided into three major parts, including keynote speeches, invited speeches and oral presentations. Over 60 leading environmental and energy researchers, engineers and scientists from numerous countries, including Malaysia, Russia, Indonesia, India, Peru, Romania, France, the UAE, the Philippines, Belgium, México, Australia, Canada, China, Thailand and Poland. Topics included environmental aspects dealing with Wastewater Treatment, Water Analysis, Hydraulic Engineering, Environmental Management, Waste Utilization and Sustainable Development. Renewable energy-related aspects were dealt with in separate sessions and covered Renewable Energy Technology and the associated Chemical Engineering and Fluid Mechanics. All presentations were oral with ample time given for Questions and Answers at the end of the presentation. One presentation per session was selected as "best presentation" for its generally outstanding quality.

All full papers presented at the 2022 4th International Conference on Environmental Sciences and Renewable Energy (ESRE 2022) are included in these Proceedings. All these papers were peer-reviewed by conference committee members and international experts, to guarantee their novelty, their quality and their research importance.

We believe that the conference was of a high and fruitful level, while meeting international standards. The conference series will continue in the future and will again provide an effective platform for further exchange of advanced know-how and knowledge, while also fostering potential international research collaborations in the topics of Environmental Sciences and Renewable Energy.

We would like to acknowledge all scientists and administrative staff who have supported ESRE 2022. Each individual and institutional support was very important for the success of this conference. Especially, we would like to thank the organizing committee for their professional organization and coordination of the peer-review of the papers.

Beijing, China\Leuven, Belgium

Jan Baeyens
General Chair, ESRE 2022

Contents

Part II Waste Management, Waste Utilization and Sustainable Development

Part III Renewable Energy Technology and Energy Chemical Engineering

Part I
Environmental Science and Ecological Protection

Chapter 1
Pollution Risk Assessment of Oil Spill Accidents in the Liao-Dong Bay of China

Guoxiang Liao, Xishan Li, Ruirui Wang, Wei Lei, Junsong Han, and Chang-an Liu

Abstract This study presents a stochastic modeling and risk assessment framework to investigate the oil spill pollution risks in the Liaodong Bay of China, where large-scale oil spill accidents from ships, ports, and offshore fields frequently occur. The oil spill risk assessment considered two basic statistic indicators: (1) the probability of pollution, and (2) the minimum arrival time. The risk assessment was carried out for environmentally sensitive areas in the Liaodong Bay area, i.e., the marine protected areas (MPAs) and the marine ranches (MRs), through overlaying and analyzing the locations with the spatial distributions of oil spill pollution. The temporal changes of pollution probability for representative MPAs and MRs over 10 d were also presented. The pollution risk assessment and mapping results of this study could provide valuable information for regional accidental oil spill emergency response planning and risk management.

Keywords Oil spill accidents · Pollution risk assessment · Probability of pollution · Minimum arrival time · Liaodong Bay

1.1 Introduction

Over the past decades, pollution and hazards from accidental oil spills have been highly concerned worldwide (Li et al. 2019). Recent statistics showed that the number of ship-related oil spills (>7 t) has been on a trend of decline (ITOPF 2020). However, once oil spill accidents happened, much more severe impacts would be posed to

G. Liao (✉) · X. Li · R. Wang · W. Lei · C. Liu
National Marine Environmental Monitoring Center, Dalian 116023, China
e-mail: liaogx2008@126.com

X. Li · C. Liu
College of Environmental Science and Engineering, Dalian Maritime University, Dalian 116026, China

J. Han
Liaoning Maritime Safety Administration, Dalian 116001, China

J. Baeyens et al. (eds.), *Proceedings of 2022 4th International Conference on Environment Sciences and Renewable Energy*, Environmental Science and Engineering, https://doi.org/10.1007/978-981-19-9440-1_1

marine ecosystems and human activities in coastal zones (Gao et al. 2019; Li et al. 2020, 2018, 2021a, b). For instance, during the 2010 *Deepwater Horizon* oil spill accident, approximately 4 million barrels of oil were spilled into the Gulf of Mexico, leading to extensive impacts on both natural resources and social economics (EPA 2020). The 2011 *Penglai 19–3* oil platform seabed oil spill accident in the Bohai Sea of China resulted in approximately 6,200 km^2 seawater and 1,600 km^2 sediments were polluted (China State Oceanic Administration (SOA) 2015). Therefore, there are still urgent needs to conduct a risk assessment, especially for those areas with high occurrence probabilities of accidents, so to adopt effective response strategies and measures to mitigate the impacts on those sensitive ecological or resources, e.g., marine protected areas, fishery resources, mariculture areas, tourism areas, coastlines, and coastal wetlands.

Oil spill risk could be defined as the combination of the probability that a particular spill event would occur and the magnitude of the consequences or impacts of the spill (Etkin et al. 2017; Amir-Heidari and Raie 2018). The occurrence probabilities of oil spills from ships or oil wells are usually estimated based on the historic accident data and the related statistic methods (ITOPF 2020; China Maritime Safety Administration (MSA) 2011; ; Det Norske Veritas (DNV) 2011). As for the consequences or impacts resulting from oil spill pollution, numerical modeling approaches have been widely used over the past decades (Reed et al. 1999; Spaulding 2017), in which the dynamic spatial distributions of spilled oil at different time and its potential impacts on the environmental sensitive receptors could be easily assessed. Traditionally, the oil spill pollution risks were assessed using the deterministic modeling approach, in which only scenarios of unfavorable wind directions and tidal current conditions were simulated (China Maritime Safety Administration (MSA) 2011; Lamine and Xiong 2013). However, it might miss some important scenarios and could not reflect the uncertainty of risk nature, leading to insufficient information for decision-makers to set up the regional response resources allocation. To address these uncertain problems, various new methodologies based on the stochastic modeling approach have been developed to provide more comprehensive and quantitative risk assessment (Amir-Heidari and Raie 2018; Skognes and Johansen 2004; French-McCay 2004; Ciappa and Costabile 2014; Lee and Jung 2015; Niu et al. 2016; Guo 2017; Liao et al. 2016; Lee et al. 2020). For instance, Skognes et al. developed a stochastic oil spill model, namely StatMap, that could statistically assess the pollution risks of the sea surface, shoreline, and water column (Skognes and Johansen 2004). Lee et al. assessed the oil spill pollution risk in Garorim Bay in Korea by using two factors: the impact likelihood and the first impact time of the spilled oil (Lee and Jung 2015). Liao et al. assessed the impacts of oil spills on a marine protected area by using both deterministic and stochastic modeling approaches (Liao et al. 2016).

The Bohai Economic Rim is an emerging economic powerhouse in Northern China. Liaodong Bay locates in the northeastern Bohai Sea and is one of the three major bays of the Bohai Sea. Although few large oil spill accidents were recorded in recent years, the potential oil spill risk is getting high due to the rapid development of large crude oil terminals, active maritime transportation, and drilling or production in offshore oil fields (Han et al. 2010; Liu et al. 2017, 2015a). Except for research on the occurrence probability of spill accidents and identification of the high-risk areas (Han et al. 2010; Liu et al. 2017, 2015a), environmental assessment of oil spill pollution risks from ports and channels construction was also reported (Liu et al. 2015a; Li et al. 2014; Zhang 2012; Wang 2017; Pang 2018). However, the pollution risk assessment was assessed by using the deterministic modeling approach in most previous studies. Besides, a comprehensive pollution risk assessment for large oil spills (>700 t) that potentially form ships and offshore fields at the Liaodong Bay regional level is not yet available.

Therefore, in the present study, we proposed a stochastic modeling and pollution risk assessment framework to further analyze the oil spill pollution risks in Liaodong Bay. The overall pollution risks to the sea areas and coastlines resulting from hypothetical large oil spill accidents at six hypothetical sites were presented. And eight national marine protected areas (MPAs) and six national marine ranches (MRs) were selected as the sensitive receptors for demonstrating the spatial and temporal characteristics of pollution risks. The results generated from this study could provide more effective regional oil spill response planning and risk management for the Liaodong Bay.

1.2 Materials and Methods

1.2.1 Study Area

General Feature and Sensitive Receptors. Liaodong Bay is one of the three major bays of the Bohai Sea in China. It locates on the west side of the Liaodong Peninsula and in the northeastern of the Bohai Sea. It is a typical semi-closed bay between the longitudes from 119°19′ E to 122°20′ E and the latitudes from 38°44′ N to 40°58′ N (Fig. 1.1). It is a shallow bay with an average depth of 22 m (Shuang et al. 2014). According to the water levels monitoring data, the tidal pattern of Liaodong Bay is regularly semidiurnal. Despite there are residual currents in the areas of shallow water or estuaries because of the topography and the monsoon, the magnitudes of residual currents are smaller than the tidal currents. Based on the field observed data from the seabed platform in the autumn of 2010, the residual current was feeble in the eastern area of Liaodong Bay and mainly directed southwestward, with a typical velocity of 3~5 cm/s, 3 cm/s, and 2~5.5 cm/s from north to south, respectively (Shuang et al. 2014). The average residual current observed during December 2015 along the west coast of central Liaodong Bay was basically in the SW direction with a magnitude of 1~10 cm/s (Shi et al. 2018). Generally, Liaodong Bay's prevailing wind directions are mainly northwesterly or northeasterly in winter. In contrast, the dominant directions are mainly southwesterly or southeasterly in summer. It is noted that the dominant wind directions vary from place to place due to the large region.

For the large amount of nutrients brought by the connected seven rivers, i.e., Daliao River, Liao River, Daling River, Liugu River, Xiaoling River, Luan River,

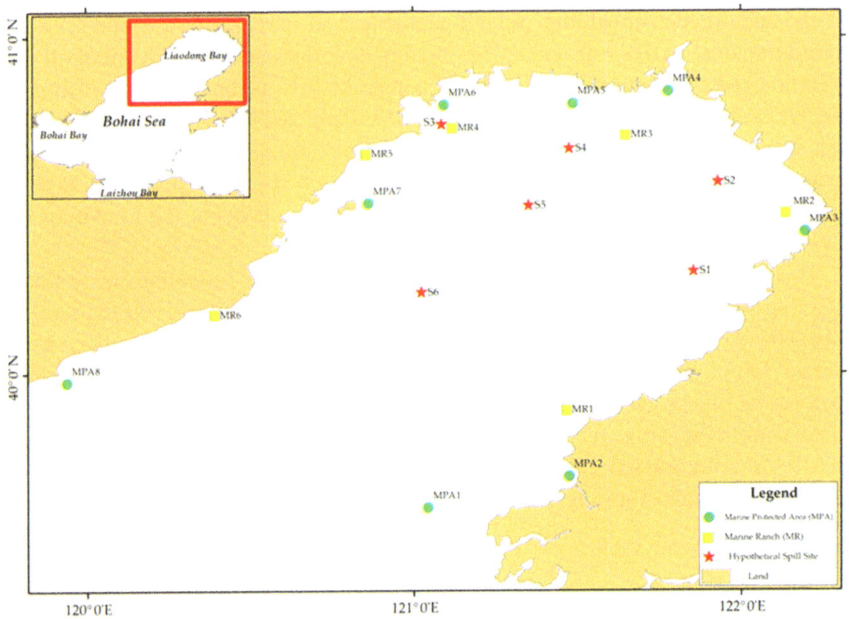

Fig. 1.1 Location, hypothetical spill sites, national marine protected areas, and national marine ranches in the Liaodong Bay of Bohai Sea

and Fuzhou River, the central sea areas of Liaodong Bay is an important spawning and feeding ground for fishes, cephalopods, and jellyfish in the Bohai Sea (Cao et al. 2017). Besides, a large area of tidal flats is distributed in the river estuaries, which are important habitats or stopovers for migratory waterbirds. Of these coastal wetlands, the Liaohe River estuary is a coastal wetland of global importance, one of the important stopovers of the East Asian—Australasian Flyway (EAAF). Recent studies also showed that the Daliao River estuary and the Beidaihe coast area were also widely recognized as important coastal wetlands of global importance (Chen et al. 2017, 2018). To protect the marine biodiversity and the important resources of Liaodong Bay, eight MPAs of national and local levels have been established in Liaodong Bay (Fig. 1.1).

To effectively use the rich nutrients and minimize the impacts of traditional fisheries, many mariculture farms were distributed in the coastal zones of Liaodong Bay. Being aware of the importance of sustainable development, both central and local governments actively promote the construction of modern marine ranches in recent years to better balance ecological conservation and human utilization. According to the construction plan of national marine ranching demonstration zones (2017–2025) published from the Ministry of Agriculture and Rural Affairs of China, a total of six pilot national MRs have been established in Liaodong Bay by 2017 (Fig. 1.1), and more MRs will be established by 2025. In addition to the marine fishery industries, transportation, and aquaculture industries, the tourism industry is also important for the administration cities along the coast of Liaodong Bay (Cao et al. 2017).

From the perspective of oil spill risk assessment, the above-mentioned marine and coastal ecological resources and human utilization activities are sensitive receptors to accidental oil spill pollution in Liaodong Bay.

Hypothetical Accidental Oil Spill Sites. Liaodong Bay represents the border separating Liaoning Province and Hebei Province. There are six municipal administration areas along the coast, i.e., Dalian, Yingkou, Panjin, Jinzhou, Huludao, and Qinhuangdao. As part of the Bohai Economic Rim, the society and economy in these areas developed rapidly over the past decades. Meanwhile, the development and expansion of ports or channels in Liaodong Bay have been remarkable since 2010. Taking the Panjin Port as an example, according to the Panjin City Government Work Reports in 2013 and 2019, the import and export of cargos increased from 2.0×10^7 t in 2012 to 7.0×10^7 t in 2018 (http://www.panjin.gov.cn/html/1404/). Besides, Liaodong Bay is rich in oil and/or gas resources under the seabed. According to the electronic navigation charts, many oil fields are distributed in Liaodong Bay (Liu et al. 2015a; Guo et al. 2019).

Compared to the large spill accidents that frequently happened in other two bays (e.g., the *Penglai 19–3 accident* near the Laizhou Bay in 2011 and *Tianjin 8.12 chemical spill accident* in Bohai Bay in 2015), very few large oil spill accidents (most are small spills in the port areas) were recorded in Liaodong Bay in history (Han et al. 2010). However, with the rapid development of ports and active maritime transportation (especially transportation of crude oils), and the aging of the undersea oil and/or gas pipelines, the occurrence probability of oil spill accidents from different sources (e.g., ships, ports, and offshore fields) are relatively high in the Liaodong Bay (Liu et al. 2015b).

Considering that Liaodong Bay of Bohai Sea is an important region for marine ecological conservation and social-economic development, the accidental oil spill pollution risks from ships, ports, and oil fields are increasing. Hence, it is necessary to conduct pollution risk assessment resulting from accidental oil spills to provide comprehensive decision-making information for emergency planning and response. In the present study, six hypothetical spill sites were selected (Fig. 1.1). Site 1, Site 2, and Site 3 were assumed as the sites that oil spill accidents that happened in the approach channels near Yingkou, Panjin, and Jinzhou Ports, while Site 4, Site 5, and Site 6 were assumed as the hypothetical spill sites from offshore oil fields in the central Liaodong Bay.

1.2.2 Stochastic Modeling and Pollution Risk Assessment

General Framework. In this study, we proposed a general framework for oil spill pollution risk assessment based on the stochastic modeling. As shown in Fig. 1.2, the framework includes three main parts: inputs (data preparation), calculation (stochastic modeling), and outputs (pollution risk assessment).

In the inputs (data preparation) part, when the study area was determined, the data and information related to the spill accidents should be collected, including: (a) the basic data of the study area, such as the geographic data the social-economic data, and so on; (b) data for determination of hypothetical oil spill scenarios, including locations and sizes of historical oil spill accidents, physical and chemical properties of spilled oil (e.g., density, viscosity, surface tension, volatility, solubility), and so on; (c) environmental sensitive receptors that might be exposed to oil spills; (d) reliable data of marine environmental conditions (e.g., temperatures, currents, winds), which could be generated from hydrodynamic and weathering prediction models.

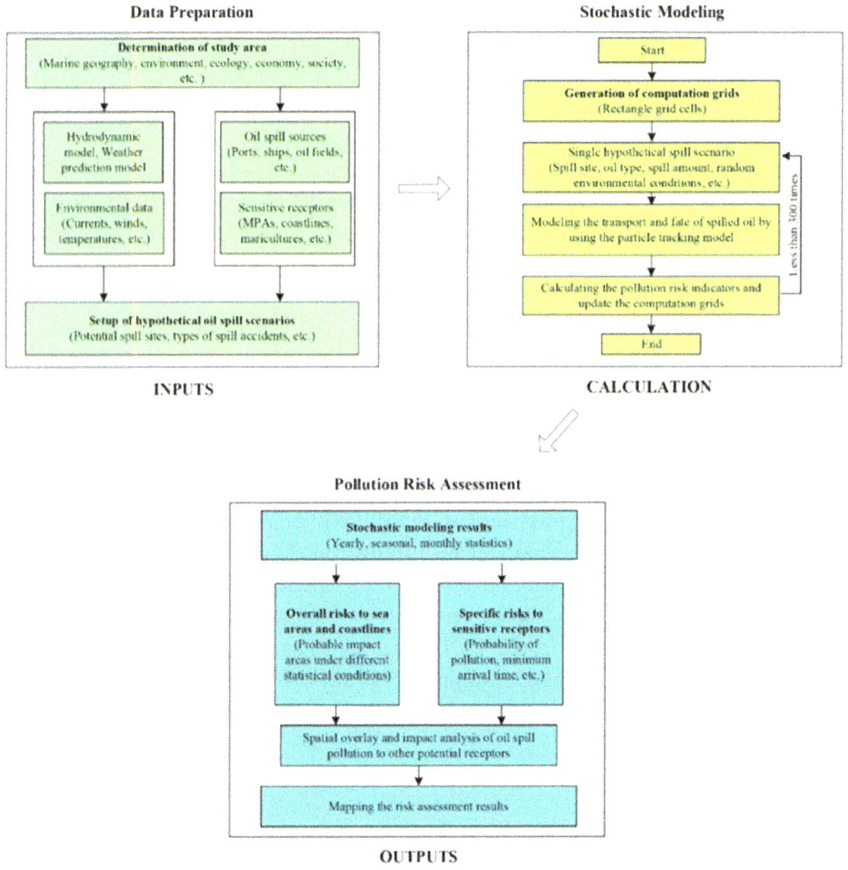

Fig. 1.2 The proposed framework for oil spill pollution risk assessment based on the stochastic modeling

When these data were prepared, randomizing the spill information and environmental conditions could generate hypothetical accidental oil spill scenarios for the study area.

The calculation (stochastic modeling) part is the core of this framework. First, the study area should be divided into a number of computing grids (e.g., rectangle grids) to facilitate the statistic analysis in the processes of calculation and outputs. Then, the stochastic modeling of the hypothetical spill scenarios for one hypothetical spill site could be conducted as follows. Each scenario would be set up with varying input parameters, including the spill sites, the spill amount and duration, the random spill time, and the environmental conditions. The transport and diffusion of spilled at sea for each scenario could be simulated using the particle tracking model. The weathering processes could also be simulated according to the model's abilities. In each run of the modeling, the pollution risk indicators of computation grids could be calculated according to each oil particle's locations and properties. The spill scenarios would repeat running hundreds of times until it reached the threshold number of stochastic modeling. As suggested by China MSA (2011), the threshold number of stochastic runs should be more than 300 (China Maritime Safety Administration (MSA) 2011).

The outputs of stochastic modeling usually are the yearly/seasonal/monthly statistic data. Based on these data, overall risks to sea areas and coastlines and specific risks to sensitive receptors could be assessed using the pollution risk indicators, e.g., the probability of pollution, minimum arrival time, thickness of surface slick, and concentration in the water column. All the risk assessment results could be presented as maps or table forms, including: (a) maps showing the probability that a specific area/coastline would be polluted by the spilled oil, expressed as a percentage (%); (b) maps showing the shortest time (hours or days) to first oil contact for a specific area/coastline segment; (c) mass balance overtime for the fate of the spilled oil (on water, evaporated, retained on the shoreline, dissolved, dispersed, or biodegraded, etc.); (d) the other statistic data according to the research/practical needs.

Hypothetical Scenarios Setup. Table 1.1 summarizes the parameters, values, and descriptions for the hypothetical oil spill scenarios in six hypothetical spill sites. As we know, there are many types of offshore oil fields-related accidents, such as surface/undersea blowouts, platform container leakage, undersea pipeline leakage, and so on (Liu et al. 2017), and ship-related spill accidents due to allision/collision, grounding, hull failure, etc. (ITOPF 2020). Considering that it is not possible to model all types of spill accidents, only two types of oil spill accidents were selected to demonstrate the ability of the stochastic modeling approach. In China, the spill size of 1000 t (and more) is treated as the highest level in national emergency response plans for ships and offshore field oil spill accidents (China State Oceanic Administration (SOA) 2015; China Ministry of Transportation (MOT) 2018). Thus, 1000 t of fuel oil or crude oil, with assuming the spill duration of 48 h, was used for all hypothetical oil spill scenarios. To get better statistical results, 450 stochastic runs for each spill site were modeled based on the environmental data ranging from 2012 to 2015. Besides, four seasons (i.e., spring, summer, autumn, and winter) were considered in this study to further analyze the seasonal variation of oil pollution risks resulting from these spill sites.

Stochastic Modeling of Oil Spills. Previous research showed that the deterministic oil spill transport and fate model was the foundation for stochastic modeling (Amir-Heidari and Raie 2018; Skognes and Johansen 2004; French-McCay 2004; Guo 2017). The oil spill transport and fate model, which has been validated in the Pearl River estuary in Southern China and Dalian Bay in Northern China, was used in the stochastic modeling in this study, and the detail could be found in our previous studies (Liao and Li 2010; Liao et al. 2010). In this model, the particle tracking method was adopted to simulate the transport and fate of spilled oil at sea. Briefly, the spilled oil would be represented by many oil particles of Lagrangian characteristics, and each particle represented a certain amount of oil mass. It would be treated as oil slicks at the sea surface and oil droplets in the water column. In each time step, each oil particle's changing properties, including its location, mass, density, and viscosity, etc., which were calculated by the algorithms for different weathering processes, would be recorded into files.

Table 1.1 Hypothetical oil spill scenarios and parameters for stochastic modeling

Parameters	Values and descriptions
Spill sites	Site 1, Site 2, and Site 3 represented the sites where hypothetical oil spill accidents happened at the approach channels of Yingkou, Panjin, and Jinzhou Ports, respectively; Site 4, Site 5, and Site 6 represented the sites where oil spilled from oil fields in the middle of Liaodong Bay
Oil type	Fuel oil 180 CST (with a density of 967.0 kg/m^3, surface tension 200 $\mu N/cm$ at 15 °C) for scenarios of spilled from Site 1, Site 2, and Site 3; Crude oil (with a density of 949.3 kg/m^3, surface tension 200 $\mu N/cm$) for scenarios of spilled from Site 4, Site 5, and Site 6
Amount and duration	The amount of spilled oil was set to 1000 t, and the duration of the spill was set to 48 h for all the scenarios
Modeling duration	10 d
Currents and winds	The sea current field data is composed of two parts, i.e., tidal current and residual current, in which the tidal current data was generated by a two-dimensional hydrodynamic model; the seasonal average residual current field data was calculated by the finite volume coastal ocean model based on long-term climate data. The wind field data was obtained from a Weather Research and Forecasting (WRF) Model, ranging from 2012 to 2015. More descriptions of current and wind data would be introduced below
Computation grids	The Bohai Sea is divided into 400 by 400 grid cells, and the resolution of each grid cell is approximately 1 km width and 1 km height
Stochastic runs	450 runs for each spill site, with random spill time and varied tidal current and wind conditions
Statistical seasons	Spring (March, April, and May), Summer (June, July, and August), Autumn (September, October, and November), Winter (January, February, and December)

In this way, the spatial distribution and changing properties of spilled oil could be simulated at each time step. For an oil particle which locates at $X_O(x_o,y_o)$, its next position could be calculated by the following equation:

$$X_N = X_O + (U_C + f \cdot D \cdot U_W) \cdot \Delta t + R\sqrt{6k_h \Delta t} \tag{1.1}$$

where X_N and X_O are the new and old positions of an oil particle, respectively; U_c is the sea surface current velocity, including the tidal current and residual current; U_w is the wind velocity above the sea surface; f is the wind drift coefficient, and it is usually assigned as 0.03; D is the transformation matrix in view of the wind deflection effect; R is a random number obeying normal distribution with mean value zero; k_h is the horizontal turbulent diffusion coefficient; Δt is the time step.

As mentioned above, the transport of spilled oil at sea is mainly governed by the surface currents and winds. Since the Liaodong Bay is an inner bay with shallow water depth, a two-dimensional hydrodynamic model was used to generate the tidal currents fields for stochastic modeling. Considering that the spilled oil will be transported long distances, the whole Bohai Sea area was selected as the model domain. The hydrodynamic model was verified with the observed water level and current speed data, and a detailed introduction could be founded in our previous works (Liao et al. 2016, 2017).

Figure 1.3 shows the snapshots of tidal current fields at different statuses, and the tidal current moves from southwest to northeast in flood tide and from northeast to southwest in ebb tide. Besides, the variation of residual current fields with four seasons was also considered. Figure 1.4 shows the average residual current fields in spring, summer, autumn, and winter.

The wind data was obtained from reanalysis data based on the numerical results outputted by the Weather Research & Forecasting Model (WRF) covering the Bohai Sea and ranging from January 1, 2012, to December 31, 2015. The sea surface wind fields are in high temporal and spatial resolution (i.e., 1 h time intervals, and a horizontal resolution of 0.1° by 0.1°), sufficient to represent spatial and time variations of the spilled oil trajectories. As shown in Fig. 1.5, the wind directions were quite

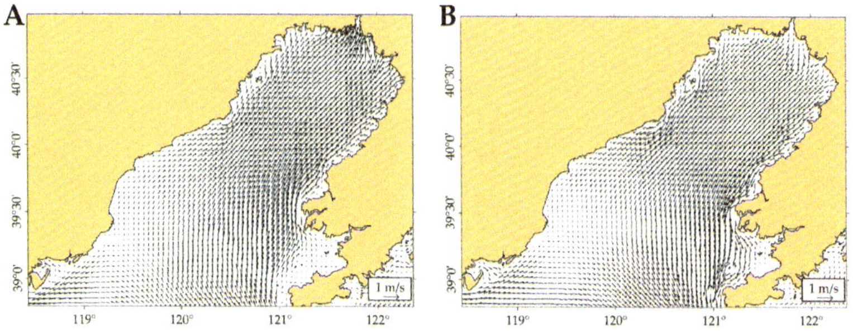

Fig. 1.3 Numerical simulated tidal current fields at different moments: **a** Flood tide, **b** Ebb tide

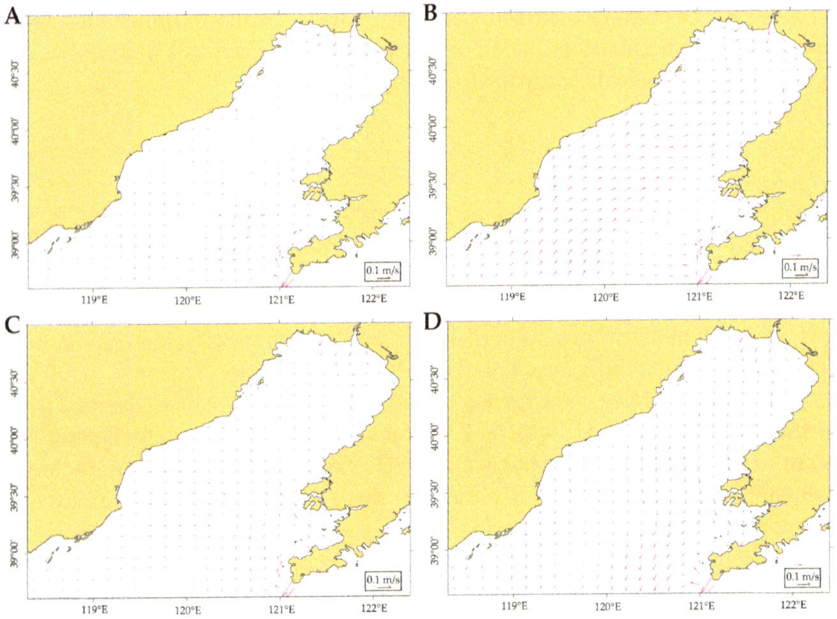

Fig. 1.4 Numerical predicted average residual current fields in four seasons: **a** Spring, **b** Summer, **c** Autumn, **d** Winter

different in the summer and winter seasons, and the wind directions also vary from place to place.

Pollution Risk Assessment. As stated in the proposed framework and previous studies, many indicators could be used for assessing the oil spill pollution risks. Of these indicators, the probability of oil pollution and minimum arrival time are the two widely accepted basic pollution risk indicators. The former indicator is important for deciding which areas to be protected in priority and can be used to plan the allocation

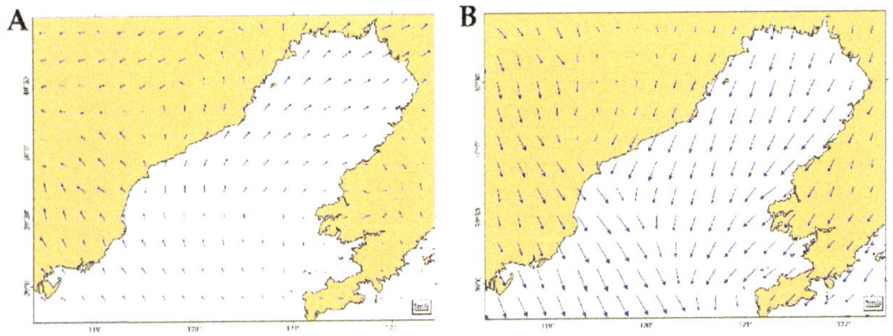

Fig. 1.5 Numerical predicted sea surface wind fields at different moments: **a** Summer, **b** Winter

of response resources. The latter indicator can provide valuable information to adopt quick and reasonable actions to mitigate oil pollution impacts.

In the present study, the stochastic modeling domain will be discretized into a large number of rectangular grids. Therefore, by combining the oil particles' locations at a given time, which is calculated by the oil spill transport and fate model, the probability of pollution and the minimum arrival time of a specific computation grid (i,j) could be calculated using Eqs. (1.2) and (1.3), respectively.

$$P_{grid}(i, j) = \frac{M(i, j)}{N} \times 100\% \tag{1.2}$$

$$T_{\min}(i, j) = \min[T(i, j)_n], n = 1, 2..., N \tag{1.3}$$

where, $P_{grid}(i,j)$ is the probability of oil pollution for the grid (i,j); $M(i,j)$ is the times of being polluted; $T_{min}(i,j)$ is the minimum arrival time that the spilled oil reached grid (i,j); $T(i,j)_n$ is the time that the spilled oil reached *grid* (i,j) in the nth stochastic run; N is the total number of stochastic runs.

1.3 Results and Discussion

1.3.1 Pollution Risks to Liaodong Bay

The pollution risk maps resulting from hypothetical large oil spill accidents at six spill sites in spring, summer, autumn, and winter were shown in Figs. 1.6 and 1.7. The seasonal dominant wind directions significantly influence the spatial distribution of oil pollution risks. In summer, the prevailing southerly wind moved the spilled oil towards the northern shoreline from spill sites. Due to the strong northerly wind with higher wind speed in winter, the spilled oil could be transported to the far southern part of Liaodong Bay, resulting in much larger impact areas than those in other seasons. While in autumn, the spatial distribution pattern of pollution probability was similar to that in spring, but the distribution of impact areas of higher probabilities was different. The reason is that the wind directions usually change from N, NW to S, SE in spring, while the wind directions change from S, SE to N, NW in autumn.

The impact areas with a pollution probability greater than 5% are listed in Table 1.2. It is interesting that the probable impact areas resulting from accidental oil spills at those sites in the western and middle of Liaodong Bay were generally larger than those on the eastern side. Due to those sites were in an open marine environment, the spilled oil was easily transported to other places without the shoreline restriction. For example, in the case of Site 5, the probable impact areas were the largest in winter, followed by the autumn and spring, and the smallest in summer, and the areas were 11,628.02, 8254.67, 7315.30, and 3392.46 km^2, respectively. While in the case of Site 1, the probable impact areas in four seasons were generally smaller

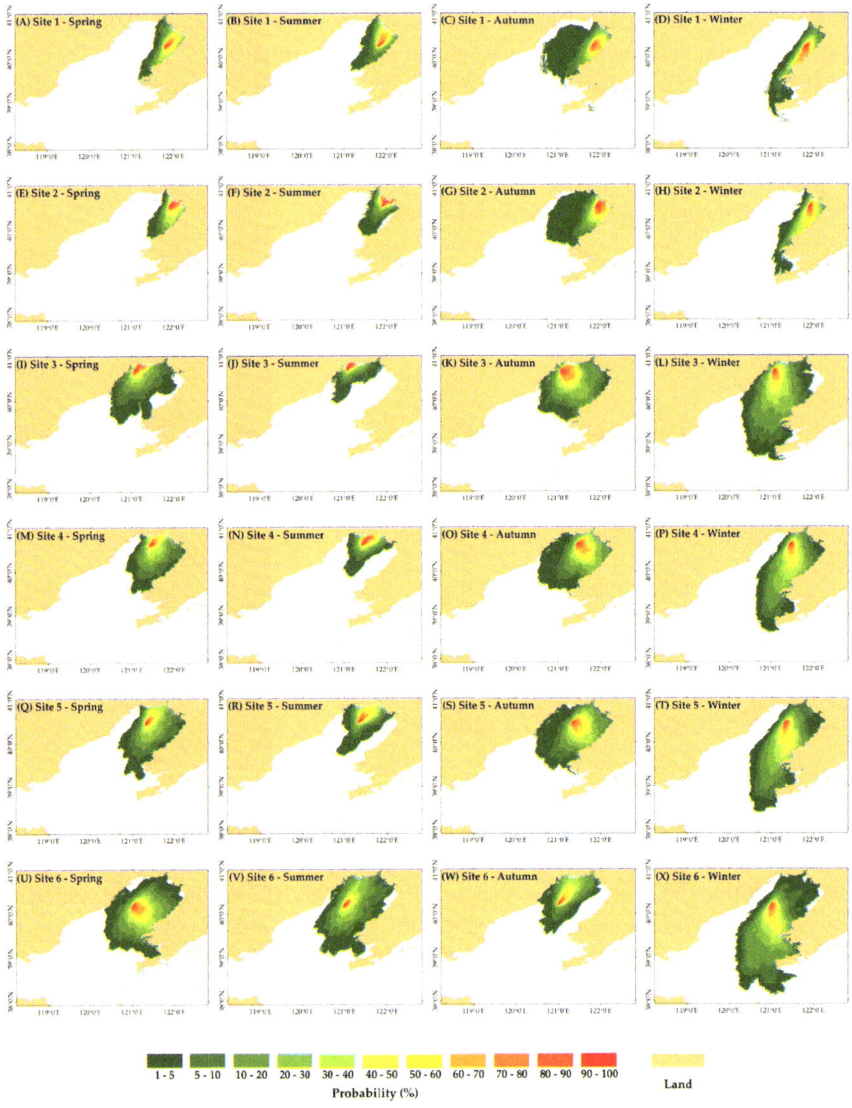

Fig. 1.6 Spatial distribution of the probability of pollution (%) to sea areas and coastlines under different seasonal conditions after spills from six hypothetical spill sites

than those of Site 5, and the largest impact areas occurred in autumn. It is well known that the minimum arrival time is critical for accidental oil spill emergency response. Figure 1.7 shows the spatial distribution of the minimum arrival time from six spill sites. Given the specific sensitive receptors' locations, the minimum arrival time can be determined from these results. A detailed example analysis of minimum arrival time would be presented in the following section.

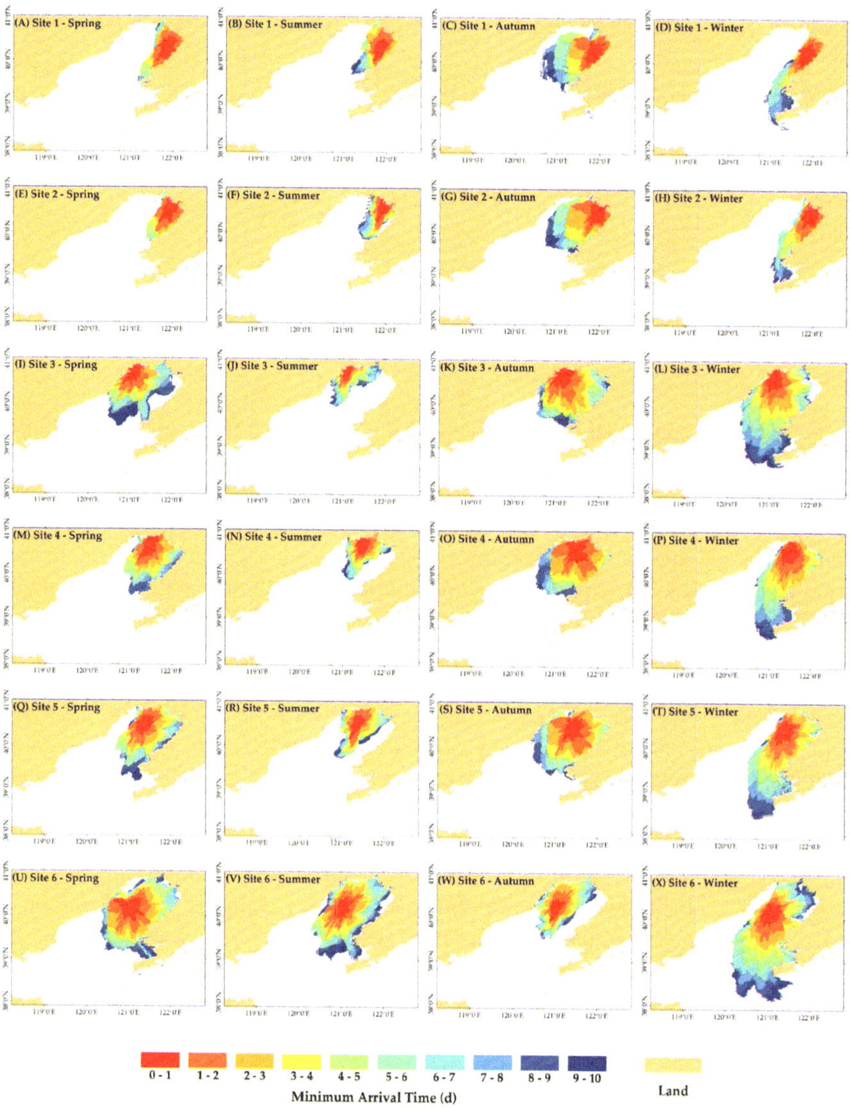

Fig. 1.7 Spatial distribution of the minimum arrival time (d) to sea areas and coastlines under different seasonal conditions after spills from six hypothetical spill sites

1.3.2 Pollution Risks to Sensitive Receptors

Based on the overall modeling results (Figs. 1.6 and 1.7), the pollution risks from six spill sources to the specific sensitive receptors could be generated by overlaying the spatial distribution of oil pollution with sensitive receptors' locations. However, it is

Table 1.2 Impact areas of more than 5% probability of oil pollution in the 10th day under different season conditions

Spill sites	Probable impact areas (km^2)			
	Spring	Summer	Autumn	Winter
Site 1	3733.62	3086.66	3617.99	3419.22
Site 2	2611.72	1973.36	3608.43	3541.54
Site 3	5799.68	1955.21	9213.16	13,336.68
Site 4	6095.92	2765.57	7237.89	10,445.92
Site 5	7315.30	3392.46	8254.67	11,628.02
Site 6	9344.08	5201.46	9593.50	14,751.00

too complex to present all the risk assessment results from the six sites; the pollution risk to the MPAs and MRs from Site 5 was selected as a case study in the present study. The sizes of MPAs vary from tens to hundreds of square kilometers (km^2), while the sizes of MRs are several square kilometers. Besides, China government is conducting the integration and optimization of MPAs at a nationwide scale. In such a context, the probability of pollution and minimum arrival time at the representative locations of the MPAs and MRs were presented and analyzed to illustrate the risks simply.

Figures 1.6 and 1.7 show the probability of pollution and minimum arrival time to the eight MPAs and six MRs in four seasons. As shown in Fig. 1.8a, the MPAs and MRs on the east side (e.g., MPA1, MPA2, MPA4, MR1, and MR3) of Liaodong Bay had high probabilities of being exposed to the pollution after oil spilled from Site 5. For the MPAs and MRs (i.e., MPA1, MPA2, and MR1) that locating at the southeast of Site 5, they had higher probabilities of being polluted in winter. While the MPAs and MRs (i.e., MPA4, MPA5, MR3) located in the north of Site 5, they had higher probabilities of being polluted in summer. Generally, the variation of wind speeds and directions under different seasons had significant impacts on the minimum arrival time. By comparing Fig. 1.8a, b, the minimum arrival time for oil transported from Site 5 to a specific receptor (e.g., MPA or MR) would be more than 240 h (i.e., 10 d) if the probability of pollution was zero. The higher probability of being polluted, the shorter it took the oil to be transported from Site 5 to the receptor.

Figure 1.8 shows the temporal changes of the probability of pollution to the MPAs and MRs after oil spilled from Sites 5 over 10 d, which indicated that the probabilities of pollution varied with the seasons and time. For example, the probability of pollution from Site 5 to MPA5 was zero on the first day due to spilled oil that did not reach to it. However, the number increased from 7.44% on the 3rd day, to 11.57% on the 5th day, and to 20.66% on the 10th day. While in summer, the probability of pollution from Site 5 to MPA5 was zero on the first day due to spilled oil that did not reach to it. However, the number increased from 8.26% on the 3rd day, to 23.97% on the 5th day, and to 42.15% on the 10th day. This information presented in Fig. 1.8 could be valuable for the emergency response and adopt effective and reasonable measures to mitigate the potential risks and make full preparations.

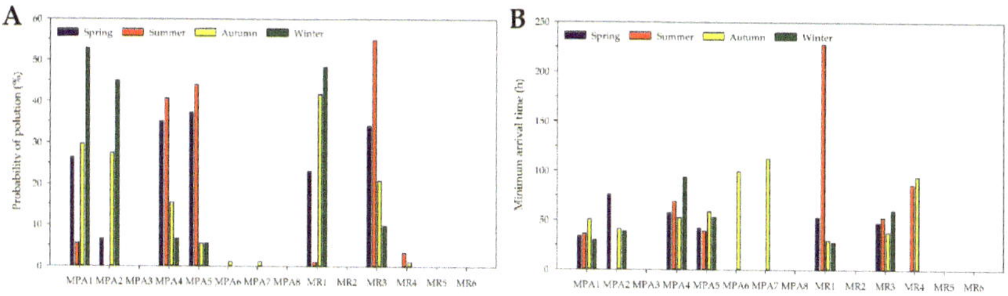

Fig. 1.8 A figure caption is always placed below the illustration. Short captions are centered, while long ones are justified. The macro button chooses the correct format automatically

Figures 1.9 and 1.10 show the temporal changes in the probability of pollution to the MPAs and MRs after oil spilled from Site 5 over 10 d, which indicated that the probabilities of pollution varied with the seasons and time. For example, in spring, the probability of pollution from Site 5 to MPA5 was zero on the first day due to spilled oil that did not reach it. However, the number increased from 6.59% on the 3rd day to 23.08% on the 5th day and 36.26% on the 10th day. While in summer, the probability of pollution from Site 5 to MPA5 was zero on the first day due to spilled oil that did not reach it. However, the number increased from 9.89% on the 3rd day to 26.37% on the 5th day and 42.86% on the 10th day. This information presented in Figs. 1.9 and 1.10 could be valuable for the emergency response and adopt effective and reasonable measures to mitigate the potential risks and make full preparations.

1.4 Conclusions

A stochastic modeling and risk assessment framework was proposed in this study. Following the basic principle of modeling, this framework consists of three main parts: inputs (data preparation), calculation (stochastic modeling), and outputs (pollution risk assessment). When the necessary data were ready, through the stochastic modeling of hundreds of hypothetical scenarios for each hypothetical spill site, the overall pollution risks to sea areas and coastlines could be quantitatively assessed with including but not limited to two basic statistic indicators, i.e., probability of pollution and minimum arrival time. The developed framework, which especially takes the seasonal variation of pollution risks into account, could be especially useful for the regional emergency response planning and decision-making in different seasons.

The Liaodong Bay of Bohai Sea, a region for marine ecological conservation and social-economic development, was selected as the study area. The hypothetical large oil spill accidents under varied environmental conditions (e.g., winds, currents) in four seasons (i.e., spring, summer, autumn, winter) were stochastically modeled using the proposed framework. The probabilities of pollution and minimum arrival time results showed that the seasonal dominant wind directions significantly impacted the spatial distribution of oil pollution risks. Besides, due to the stronger variation of wind

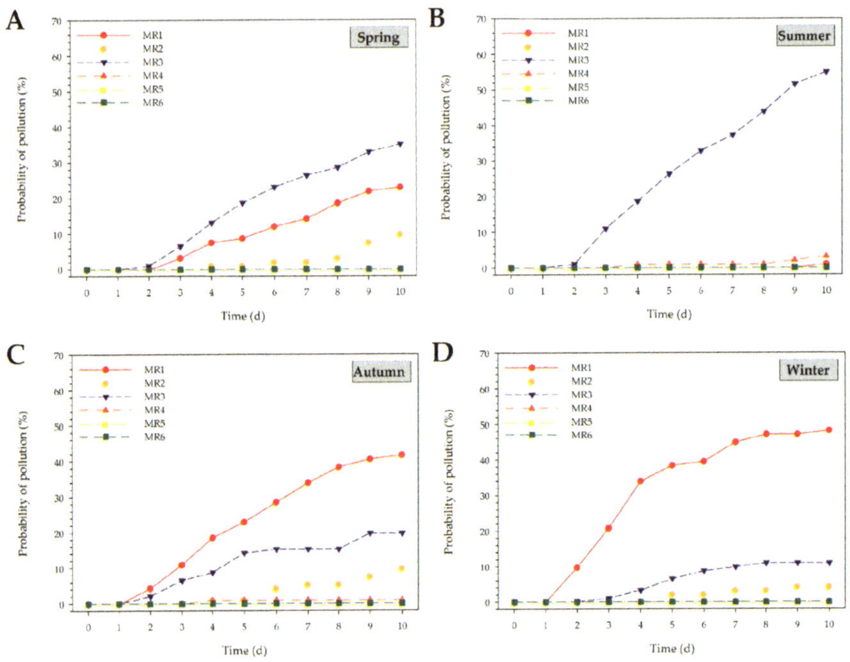

Fig. 1.9 Temporal changes of oil pollution probability at the representative locations of national marine ranches (MRs) over 10 d after an oil spill from Site 5

directions and higher wind speeds, the spilled oil's probable impact area in winter was the largest compared to the other seasons, followed by the autumn and spring, and the smallest in summer. In addition, the residual currents also have impacts on the spatial distribution of pollution risks (i.e., probabilities of pollution and minimum arrival time), especially in those areas where seasonal residual currents are relatively strong, e.g., Liaohe River estuary, Jinzhou Bay (see Fig. 1.4). The specific pollution risks to sensitive receptors, i.e., the national MPAs and MRs in Liaodong Bay, were assessed in two ways, including the spatial overlay analysis and temporal changes of pollution probabilities over time, demonstrating the excellent application ability of the proposed framework.

In summary, this study's results could help governmental decision-making once oil spill accidents happened in Liaodong Bay, especially for response planning, response resource allocations, and emergency monitoring. Further research should be conducted on the ecological risk assessment on marine mammals, migratory waterbirds, and their habitats in coastal wetlands that are vulnerable to accidental oil spill pollution, which might provide more insightful decision-making information for regional marine oil spill preparation and contingency response.

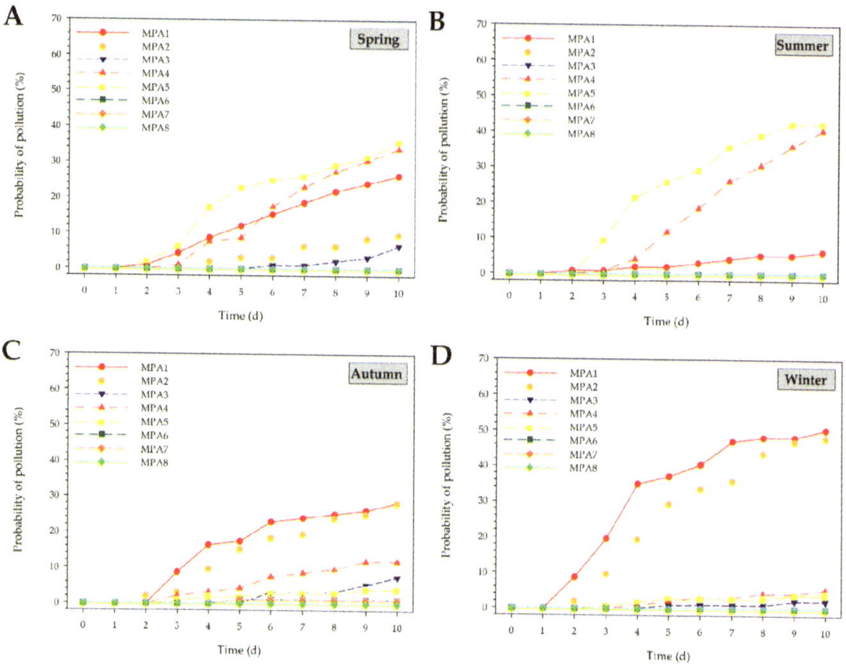

Fig. 1.10 Temporal changes of oil pollution probability at the representative locations of national marine protected areas (MPAs) over 10 d after an oil spill from Site 5

Acknowledgements This research was supported by the National Key Research and Development Program of China, Grant Number 2018YFD0900606, and the National Natural Science Foundation of China, Grant Number 42076215, 41306099, and 41806187. The authors are grateful to the editor and the anonymous reviewers for their availability to review this work. Their valuable comments and suggestions have improved the quality of the manuscript. Besides, the authors give a very special thanks to Qian Zhao (from National Marine Environmental Monitoring Center, Dalian, China) for providing the residual current data in Liaodong Bay.

References

Amir-Heidari P, Raie M (2018) Probabilistic risk assessment of oil spill from offshore oil wells in Persian Gulf. Mar Pollut Bull 136:291–299

Cao K, Xu Y, Cai Y (2017) Sensitivity assessment on the ecological environment in coastal zones based on oil spill risk taking the Liaodong Gulf as an example. J Ocean Technol 36:73–78

Chen K, Lv Y, Chen Q (2017) Yellow Sea-Bohai region coordinated waterbird survey 2017 (summary report). Wetlands International, Beijing, China

Chen K, Lv Y, Chen Q (2018) Yellow Sea-Bohai region coordinated waterbird survey 2018 (Summary Report). Wetl Int Cormor Res Group Bull: Beijing, China

China Maritime Safety Administration (MSA) (2011) Technical specification for risk assessment of marine environmental pollution by ships (Trial). China Maritime Safety Administration, Beijing, China

China Ministry of Transportation (MOT) (2018) National emergency plan for the treatment of major marine oil spills. China Ministry of Transportation, Beijing, China

China State Oceanic Administration (SOA) (2015) SOA oil spill contingency plan for offshore oil exploration and development. China State Oceanic Administration, Beijing, China

Ciappa A, Costabile S (2014) Oil spill hazard assessment using a reverse trajectory method for the Egadi marine protected area (Central Mediterranean Sea). Mar Pollut Bull 84:44–55

Det Norske Veritas (DNV) (2011) Report for Australian Maritime Safety Authority: Model of offshore oil spill risks. Managing Risk, Norway

EPA (2020) Deepwater horizon—BP Gulf of Mexico oil spill. United States Environmental Protection Agency

Etkin DS, French McCay D, Horn M, Landquist H, Hassellöv IM, Wolford AJ (2017) Chapter 2—quantification of oil spill risk. Oil Spill Technol (Second Edition) 71–183

French-McCay DP (2004) Oil spill impact modeling: development and validation. Environ Toxicol Chem 23:2441–2456

Gao Y, Xiong D, Qi Z, Li X, Ju Z, Zhuang X (2019) Distribution of polycyclic aromatic hydrocarbons in sunken oils in the presence of chemical dispersant and sediment. J Mar Sci Eng 7:282

Guo W (2017) Development of a statistical oil spill model for risk assessment. Environ Pollut 230:945–953

Guo W, Zhang S, Wu G (2019) Quantitative oil spill risk from offshore fields in the Bohai Sea, China. Sci Total Environ 688:494–504

Han J, Xiong D, Liao G (2010) Identification of high-risk areas of marine pollution accidents in the Bohai Sea. J Navig China 33:85–89

ITOPF (2020) Oil tanker spill statistics 2019. London, UK, p 20

Lamine S, Xiong D (2013) Guinean environmental impact potential risks assessment of oil spills simulation. Ocean Eng 66:44–57

Lee M, Jung J-Y (2015) Pollution risk assessment of oil spill accidents in Garorim Bay of Korea. Mar Pollut Bull 100:297–303

Lee K-H, Kim T-G, Cho Y-H (2020) Influence of tidal current, wind, and wave in Hebei Spirit oil spill modeling. J Mar Sci Eng 8:69

Li H, Lou A, Wang J, Sun X (2014) A numerical simulation of the oil spill accident at Penglai 19–3 oil field. Mar Sci 38:70–77

Li X, Ding G, Xiong Y, Ma X, Fan Y, Xiong D (2018) Toxicity of water-accommodated fractions (WAF), chemically enhanced WAF (CEWAF) of Oman crude oil and dispersant to early-life stages of zebrafish (Danio rerio). Bull Environ Contam Toxicol 101:314–319

Li X, Xiong D, Ding G, Fan Y, Ma X, Wang C, Xiong Y, Jiang X (2019) Exposure to water-accommodated fractions of two different crude oils alters morphology, cardiac function and swim bladder development in early-life stages of zebrafish. Chemosphere 235:423–433

Li X, Liao G, Ju Z, Wang C, Li N, Xiong D, Zhang Y (2020) Antioxidant response and oxidative stress in the respiratory tree of sea cucumber (*Apostichopus japonicus*) following exposure to crude oil and chemical dispersant. J Mar Sci Eng 8:547

Li X, Wang C, Li N, Gao Y, Ju Z, Liao G, Xiong D (2021a) Combined effects of elevated temperature and crude oil pollution on oxidative stress and apoptosis in sea cucumber (*Apostichopus japonicus*, Selenka). Int J Environ Res Public Health 18:801

Li X, Xiong D, Ju Z, Xiong Y, Ding G, Liao G (2021b) Phenotypic and transcriptomic consequences in zebrafish early-life stages following exposure to crude oil and chemical dispersant at sublethal concentrations. Sci Total Environ 763:143053

Liao G, Han J, Xiong D (2010) Marine oil spill transport and fate numerical modeling for complex leaking modes. J Dalian Mar Univ 36:86–90

Liao G, Dai L, Lu W, Ye J, Liu C (2016) Risk analysis of marine oil spill pollution using a stochastic simulation approach: a case study of Binzhou chenier plain and national wetland nature reserve. Mar Sci Bull 35:467–479

Liao G, Ye J, Han J, Liu C (2017) Predictive assessment of pollution and hazard resulting from marine oil spill accidents with stochastic and deterministic scenario simulations. Mar Environ Sci 36:266–273

Liao G, Li M (2010) Numerical prediction of oil spill trajectory and fate in Pearl River estuary. In: 2010 international conference on mechanic automation and control engineering, pp 2153–2157

Liu X, Meng R, Xing Q, Lou M, Chao H, Bing L (2015a) Assessing oil spill risk in the Chinese Bohai Sea: a case study for both ship and platform related oil spills. Ocean Coast Manage 108:140–146

Liu X, Guo J, Guo M, Hu X, Tang C, Wang C, Xing Q (2015b) Modelling of oil spill trajectory for 2011 Penglai 19–3 coastal drilling field, China. Appl Math Model 39:5331–5340

Liu B, Wei W, Duan M, An W, Jin W (2017) Oil spill risk assessment of offshore platform in Bohai Sea. Mar Environ Sci 36:15–20

Niu H, Li P, Yang R, Wu Y, Lee K (2016) Effects of chemical dispersant and seasonal conditions on the fate of spilled oil - modelling of a hypothetical spill near Saint John, NB. Water Qual Res J Can 51:233–245

Pang X (2018) Research on the risk and prevention of ship-source oil spill pollution at Xianrendao Port of Yingkou. Dalian Maritime University

Reed M, Johansen O, Brandvik PJ, Daling P, Lewis A, Fiocco R, Mackay D, Prentki R (1999) Oil spill modeling towards the close of the 20th century: overview of the state of the art. Spill Sci Technol Bull 5:3–16

Shi W, Xing C, Ma Y, Chen Y, Hu Z, Hou F (2018) Analysis of current observation in shallow inshore waters along the west coast of central Liaodong Bay in winter. Mar Sci Bull 37:389–395

Shuang Y, Kun W, Lun S, Guiying L, Yonggang S (2014) Statistical analysis and evaluation of nutrient distribution in Liaodong Bay. Hebei Fisheries 14–18

Skognes K, Johansen O (2004) Statmap-a 3-dimensional model for oil spill risk assessment. Environ Model Softw 19:727–737

Spaulding ML (2017) State of the art review and future directions in oil spill modeling. Mar Pollut Bull 115:7–19

Wang W (2017) A study on risk of ship pollution and preventive measures of Panjin Port crude oil and liquid chemicals berth. Dalian Maritime University

Zhang H (2012) A study on pollution risk assessment and emergency strategy on peroleum and chemical docks and ships in circumjacent sea area of 301# spot of 120,000 dwt crude oil terminal in Jinzhou Port. Dalian Maritime University

Chapter 2
Research on Ecosystem Status Evaluation of Open-Pit Mines

Fengmin Wu, Zhipeng Zheng, Xiaoye Zhang, Xiaolong Chen, and Zhong Zheng

Abstract The evaluation of ecological restoration of mines is an important part of ecological environment protection. In this paper, we first constructed the evaluation system suitable for open-pit mines in mountainous areas and introduced four factors to calculate the ecological index from biological richness, vegetation coverage, land stress and landscape pattern. Then we obtained the weights of factors by expert evaluation method and classified the ecological index into five levels using the Jenks natural breaking method. Taking Zhongliangshan of Chongqing as a study area, we evaluated the ecosystem status of 70 open-pit mines in the process of ecological restoration. The result showed, ecological environment of open-pit mine was totally at a low level compared to other regions with max ecological index of 30.38. The lowest and lower level of ecological environment accounted for more than half of the total area of open-pit mines where some mines seemed to be exposed rocks with high and steep slopes which need to be restored continually.

Keywords Ecosystem status · Evaluation · Open-pit mine

2.1 Introduction

Mining activities benefitted economic development as well as brought a series of environmental issues such as vegetation destruction, soil erosion, ground subsidence and landslides (Tan et al. 2004; Zhou et al. 2008). The purpose of environment restoration of mines was to increase green area of mines and maintain a stable geological environment through combining vegetation restoration and land reclamation (Tang et al. 2020; Zhang et al. 2016). But how to evaluate the ecosystem status accurately and

F. Wu (✉) · Z. Zheng · X. Zhang · X. Chen · Z. Zheng
Chongqing Geomatics and Remote Sensing Center, Chongqing, China
e-mail: wufengmin@dl023.net

Evaluation and Early Warning of Territorial Spatial Planning Implementation, Ministry of Natural Resources, Key Laboratory of Monitoring, Beijing, China

J. Baeyens et al. (eds.), *Proceedings of 2022 4th International Conference on Environment Sciences and Renewable Energy*, Environmental Science and Engineering, https://doi.org/10.1007/978-981-19-9440-1_2

quickly had become an urgent problem to be solved at present (Jianping 2021). Traditional methods of ecosystem status evaluation included Fuzzy Evaluation Method, Analytic Hierarchy Process and PSR model. The Technical Criterion for Ecosystem Status Evaluation published in 2015 provided a standard for the evaluation of the ecological environment and the environment condition index was widely used in cities, deserts, loess erosion areas and nature reserves (Guoqiang 2017; Xu et al. 2010; Geng et al. 2008; Lili et al. 2013). However, the research on open pit mines mainly focuses on the relationship between vegetation restoration and soil factors, surface cover, land use structure and landscape patterns (Jianping 2021). There were few studies on ecological environment assessment in mining areas.

Chongqing was a typical mountain area with small and disperse open-pit mines widely distributed. We choose Zhongliangshan of Chongqing as a study area. We firstly constructed the evaluation system and reset the indicator weights according to the actual situation in Chongqing. Since open-pit mines of Chongqing were almost sand and quarry mines, pollutants were not taking into consideration. Four factors were introduced to obtain the ecological index such as biological richness index, vegetation coverage index, land stress index and landscape pattern index in this paper based on the remote sensing data. The ecosystem status was divided into five grades by the Jenks natural breaking method to nalyse the Ecosystem Status of open-pit mines under the process of restoration. This study could provide scientific basis and theoretical support for the ecological environment management of the mining area.

2.2 Methodology

2.2.1 Study Area and Used Data

The study area was located in west Chongqing in China (Fig. 2.1). The longitude and latitude were 106°23′20″-106°24′25″, 29°26′10″-29°31′50″. It spanned three regions of Shapingba, Jiulongpo and Dadukou District, covering an area of about 28.63 km^2. There were 70 open-pit mines widely distributed with total area of 191.36 ha which included the biggest mine (60.98 ha) in the southwest of study area. The landforms types were low mountains and hills with high vegetation coverage.

The data used in this study included remote sensing data, meteorological data, soil data, DEM (digital elevation model) data, land cover data and so on. GF-1 data was applied to obtain NDVI (Normalized Difference Vegetation Index) of 2020. Soil data and vegetation type data were both produced by Southwest University. The meteorological data such as monthly rainfall was from Chongqing Meteorological Service (http://cq.cma.gov.cn/) in 2020. Soil texture data was from Chongqing Forestry Bureau. The open-pit mines were from Chongqing Institute of Geology and Mineral Resource. The landcover data was calculated from Geographical Conditions Monitoring Data of 2020. The land cover data, GF-1 data, 1:5000 DEM data and

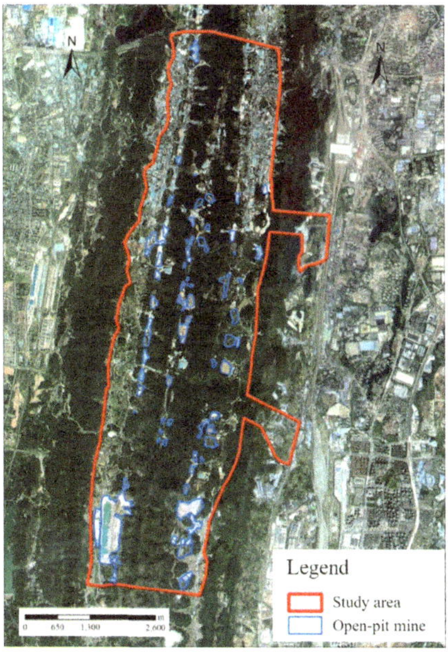

Fig. 2.1 Study area of Zhongliangshan in Chongqing

other materials were from Chongqing Geomatics and Remote Sensing Center. To unify the scales, all the data were resampled to 8 m and converted to the National Geodetic 2000 coordinate system.

2.2.2 *Biological Richness*

Biological richness was determined habitat quality when there was no dynamic update of biodiversity. The biological richness was recommended from Technical Criterion for Ecosystem Status Evaluation published by Ministry of Ecology and Environment of the People's Republic of China in 2015. Habitat quality index (H) could be obtained from landcover data calculated from:

$$H = A_{bio} \times \begin{pmatrix} 0.35 \times A_{for} + 0.21 \times A_{gra} + 0.28 \times A_{wat} + 0.01 \times A_{farm} + \\ 0.04 \times A_{constr} + 0.01 \times A_{oth} \end{pmatrix} \quad (2.1)$$

where, A_{for}, A_{gra}, A_{wat}, A_{farm}, A_{constr}, A_{oth} represents the area of forest, grass, water, farmland, construction land and other area, separately. The construction land included Building, Road, Structure, Bulldozing area. A_{bio} is the normalization coefficient of habitat quality index with the value of 511.26.

2.2.3 Vegetation Coverage

We choose the GF-1 PMS1 data (8 m) with little cloud coverage to calculate the NDVI data in June 26, 2020. The soft of ENVI 5.3 was applied to do the data preprocessing. The radiometric calibration parameters were released by China Centre for Resources Satellite Data and Application (http://www.cresda.com/CN/index.shtml). We got the mean elevation value of GF-1 data (0.4264 m) according to the 1:5000 DEM. The Flaash Atmospheric Correction model and RPC orthorectification workflow tools were used for atmospheric correction and geometric correction. The NDVI was calculated by:

$$NDVI = (B_5 - B_4)/(B_5 + B_4) \tag{2.2}$$

where, B_4, B_5 was the reflectivity in red and near-infrared bands of GF-1 data. Since the value range of NDVI was [-1.0, 1.0], the study dealt with outliers off this value range. The vegetation coverage index (V) could be calculated by:

$$V = A_{veg} \times \left(\sum_{i=1}^{n} P_i/n\right) \tag{2.3}$$

where, P_i is the NDVI of each pixel, n is the number of pixels, A_{veg} is the normalization coefficient with value of 0.012.

2.2.4 Land Stress

The land stress index (I) was calculated from different degrees of soil erosion. The soil erosion sensitivity was calculated from the Universal Soil Loss Equation (USLE) as follows:

$$I = R \times K \times LS \times C \tag{2.4}$$

where, I is the annual average soil loss per unit area, R is rainfall erosivity factor, K is soil texture factor, LS is slope and length factor computed by DEM data, C is vegetation type factor. We got the rainfall erosivity factor calculated as:

$$R = \sum_{i=1}^{12} (0.3046P_i - 2.6398) \tag{2.5}$$

where, P is monthly rainfall. The assignment of each factor on soil erosion was listed in Table 2.1.

We divided the rainfall erosivity factor (R) into six levels using natural breaks method. Finally, the land stress index was obtained from extremely sensitive (E_{exs}), sensitive (E_s), medium sensitive (E_m), construction land, other areas (E_o). The

Table 2.1 Classification of soil erosion

Soil erosion level	Rainfall erosivity(mm)	Slope and length(m)	Soil texture	Vegetation type
Slight	0 ~ 25	0 ~ 20	gravel, sand	water, herbs
Mild	25 ~ 100	20 ~ 50	coarse sand, fine sand, clay	arbor, shrub, grass
Medium	100 ~ 400	50 ~ 100	surface sand, loam	shrub, two crops/y
Strong	400 ~ 500	100 ~ 200	silt clay, loam clay	one crop/y
Very strong	500 ~ 600	200 ~ 300	sand loam	Bare ground
Extreme	> 600	> 300	sand silt, silt	non vegetation

calculation formula was as follows:

$$I = E_{ero} \times (0.4 \times E_{exs} + 0.3 \times E_s + 0.2 \times E_m + 0.05 \times E_l + 0.05 \times E_o) \tag{2.6}$$

where, E_{ero} is normalization coefficient of land stress index, the value is 236.044. E_l is calculated from high resolution remote sensing image of 2020.

2.2.5 *Landscape Pattern*

Shannon diversity index ($SHDI$) reflected the level of diversity in the landscape, which was calculated as follow:

$$SHDI = -\sum_{i-1}^{m} [P_i \ln(P_i)] \tag{2.7}$$

where, P_i is the ratio of the landscape area occupied by the i patch. In this paper, the type of landscape data was from landcover of Geographic national conditions and monitoring of 2020. The moving window algorithm of Fragstats software was applied to calculate the Shannon diversity index. A moving window of 30 m could be used to retain the gradient characteristics of spatial pattern of landscape by many times of trials.

2.2.6 *Ecological Index*

The ecological index consisted of biological richness, vegetation coverage, land stress and landscape pattern index, which was calculated as follow:

$$EI = a_1 \times H + a_2 \times V + a_3 \times I + a_4 \times SHDI \tag{2.8}$$

where, a_1, a_2, a_3, a_4 is the weight of each parameter. According to expert evaluation method, the value from a_1 to a_4 is 0.35, 0.25, 0.25 and 0.15, respectively.

2.3 Ecosystem Status Evaluation

The four factors of open-pit mines were calculated from the formulas above (Fig. 2.2). It was concluded that, the land cover types of open-pit mines included 8 first-level categories, 22 s-level categories, such as farmland, forest, grass, building, et al. Bulldozing area and Structure were the maximum composition of open-pit mines (109.91 ha) accounting for 57.43%. The farmland, grass and forest were 41.07 ha, accounting for 21.46%. The ecological restoration work of open-pit mines had not been completed yet as there was more than half area with no vegetation.

The Shannon diversity index of most open-pit mines was less than 0.3. The max value of was 2.41 which seemed lower than other areas outside. It was showed, landscape diversity and richness were both at a low level because there were a few types of patches in an open-pit mine.

The NDVI obtained from GF-1 data had highly sensitive response of vegetation index. The vegetation was in good situation in Zhongliangshan and most open-pit mines had high NDVI values. The area of NDVI > 0.4 reached 94.84 ha, counting for 49.56% of the total area of mines. The area of medium vegetation index ($0.3 < \text{NDVI} \le 0.4$) was 29.48 ha, accounting for 15.41%.

Soil erosion was very common in open-pit mines, but the proportion of soil erosion area of each mine was almost less than 10%. Soil erosion of open-pit mines were mainly under the medium level. The area of strong, very strong and extreme level was 43.92 ha, accounting for 22.95% The extreme level was distributed in southwest and almost concentrated in several open-pit mines.

We divided the ecological index into five grades: excellent, good, medium, poor and worst by the Jenks natural breaking method (Fig. 2.3).

It was showed that, the value range of ecological index was from 10.85 to 30.80 which was significantly lower compared with other areas. The area with "poor" grades accounted for 62.72% of the total area of open-pit mines and the "good" grades accounted for 15.42%. The ecological environment of open-pit mines was characterized by a high level in the northeast and a low level in the southwest. The open-pit mines with the "excellent" and "good" ecological index were mostly distributed in the forests and grasses with better vegetation coverage and less impact from human activities. It was necessary to speed up the restoration management and improve the quality of the ecological environment of open-pit mines in Zhongliangshan area.

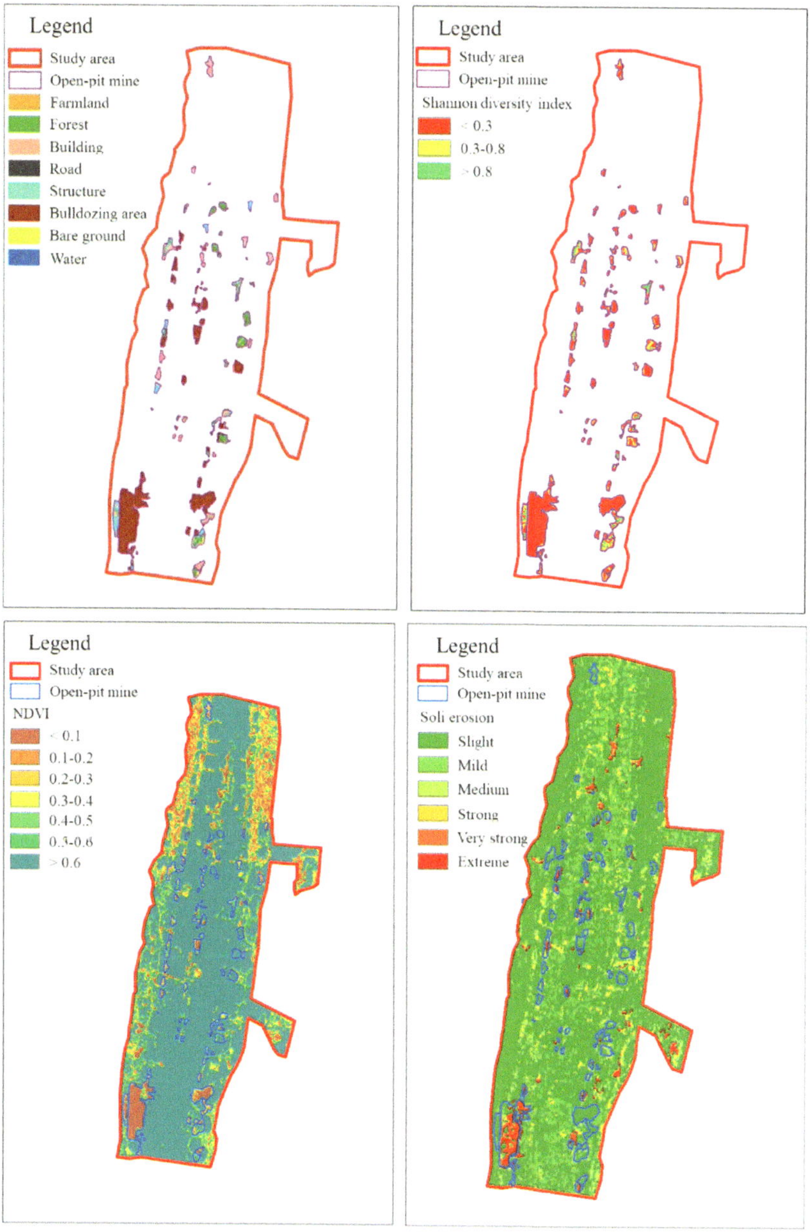

Fig. 2.2 Distributions of ecosystem factors

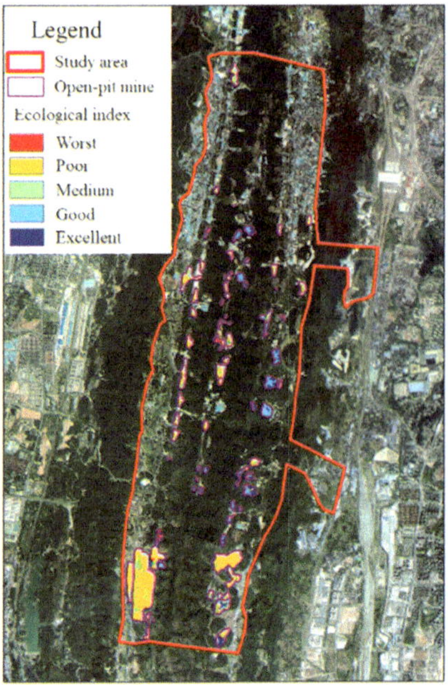

Fig. 2.3 The ecological index of open-pit mines

2.4 Conclusion

This paper introduced ecological index to evaluate the ecosystem status of open-pit mines. The landcover data from Geographical Conditions Monitoring Data of 2020 was applied to calculate the biological richness index. The vegetation coverage index was obtained by NDVI from GF-1 data. We got land stress index by soil erosion status with different weights. Landscape pattern index could be obtained from Shannon diversity. The ecological index showed there were more than half area of open-pit mines under the "poor" situation. Although many of them were in the process of ecological restoration, the open-pit mines reflected that the ecological environment were not as good as expected which should be restored as quickly as possible.

Acknowledgements This work is mainly supported by Research on application of thermal pollution monitoring of key reaches of Yangtze River based on multi-source thermal infrared remote sensing technology (No. KJQN202103410).

References

Geng L, Rutian B, Yi C (2008) On the economic benefits of land reclamation in open cast coal mine dump site[J]. China Coal 8:110–112

Guoqiang C (2017) Thesis submitted to Tianjin university of technology for the master's degree[C]. Tianjin Univ Technol, p 1

Jianping Y (2021) Ecological environment evaluation based on remote sensing ecological index in Pingshuo Open-pit Mine[J]. Open-Pit Min Technol 36(1):45–47

Lili M, Shufang T, Na W (2013) Ecological environment evaluation of the mining area based on AHP and fuzzy mathematics[J]. Remote Sens Land & Resour 25(3):165–170

Tan LG, Lu SM, Wang BW et al (2004) Mine ecological environment destruction and ecological restoration: Taking the mines of LU'AN as an example[J]. J West Anhui Univ 20(2):45–48

Tang Y, Wang LJ, Li FY (2020) Research on mine ecologic environment investment and restoration government based on "High Resolution + "[J]. Land Res Information 5:19–24

Xu C, Yuanying G, Zhongke B (2010) Ecological carrying capacity assessment of large-scale open coal mines in loess zones-A case study of Antaibao Opencast Mine in Pingshuo[J]. Chin J Eco-Agric 2:204–209

Zhang Z et al (2016) Remote sensing survey and analysis of mine geologic environment in eastern Hubei Provision under the perspective of ecological civilization[J]. Geol Surv China 3(5):21–27

Zhou YZ et al (2008) Geochemical migration model of heavy metals elements in eco-environmental system of sulfide-bearing metallic mines in South China-with specific discussion on Dabaoshan Fe-Cu-polymetallic mine, Guangdong Provision[J]. Earth Sci Front 15(5):248–255

Chapter 3
Antioxidant Response in the Respiratory Tree of Sea Cucumber *Apostichopus Japonicas* Following Acute Exposure to Merey Crude Oil

Zhonglei Ju, Xishan Li, Yuhang Zou, Wei Yang, Nan Li, Guoxiang Liao, and Deqi Xiong

Abstract Sea cucumber *Apostichopus japonicas* is mainly cultured in coastal areas with a potential marine oil spill threat. However, information about the impact of crude oil on sea cucumber is not enough. Thus, the present study aimed at investigating the antioxidant response in the respiratory tree of sea cucumber following the acute exposure (24 h, 48 h, and 72 h) to water-accommodated fractions (WAF) of Merey crude oil. Results revealed that WAF exposure caused the most significant increase in the superoxide dismutase activity at 72 h. On the contrary, an obvious decline in reduced glutathione (GSH) content was observed during all exposure time, and high concentration treatment of WAF resulted in a more severe decrease in GSH content. Moreover, we found that malondialdehyde content was illustrated in a concentration-dependent manner with WAF exposure solution, indicating that Merey crude oil exposure could induce severe lipid peroxidation in the respiratory tree of sea cucumber. Collectively, our findings provide insights into the toxic effects of crude oil acute exposure on *Apostichopus japonicas* associated with antioxidant response.

Keywords Oil Spill · *Apostichopus japonicas* · Acute toxicity · Antioxidant response

---

Z. Ju · X. Li · Y. Zou · W. Yang · N. Li · D. Xiong (✉)
Dalian Maritime University, Dalian 116026, China
e-mail: xiongdq@dlmu.edu.cn

Z. Ju · X. Li · Y. Zou · G. Liao
National Marine Environmental Monitoring Center, Dalian 116023, China

J. Baeyens et al. (eds.), *Proceedings of 2022 4th International Conference on Environment Sciences and Renewable Energy*, Environmental Science and Engineering, https://doi.org/10.1007/978-981-19-9440-1_3

3.1 Introduction

The global growing consumption of petroleum and its refined products presents oil spill threats, especially in marine ecosystems (Colvin et al. 2020). The release during oil exploitation, transport, and processing is one of the main sources of marine oil pollution, as is evidently proved a potential risk to marine organism (Yuewen and Adzigbli 2018). Sea cucumber *Apostichopus japonicas* is a typical temperate echinoderm species, widely distributed along the Asian coast from 35° N to 44° N (Hamel and Mercier 2008). *Apostichopus japonicas* is traditionally considered as one of the most precious seafoods in China for its nutritional value and curative properties. Currently, the sea cucumber is mainly cultured in outdoor ponds close to the coast and shadow sea areas, where are easily exposed to multiple environmental contaminants caused by human activities, especially from maritime shipping, sewage discharging, and oil spills (Li et al. 2020). In 2010, Dalian (China) oil pipeline explosion accident caused more than 1500 tonnes crude oil into the Dalian Bay, resulting in severe damages to marine aquaculture in Liaoning Province, and inevitably posing a serious threat to local ecosystem balance of Yellow Sea. Additionally, 2011 Penglai 19–3 oil spill in Bohai Bay polluted 840 km^2 of clean coastal water degraded, causing economic losses of more than CNY12.56 billion (Pan et al. 2015). Among the affected marine agriculture species, sea cucumber experienced a reduction of 52% in culture area in Yantai region because of the degraded costal water contaminated by oil spill (Pan et al. 2015).

A body of studies have provided evidence on the toxic effects of oil pollution on marine organism (e.g., oceanic plankton, pelagic fish, and benthos) (Beyer et al. 2016; Pasparakis et al. 2019; Li et al. 2021a, 2018). However, to the best of our knowledge, information about the effect of oil pollution on sea cucumber is still not enough. Sea cucumbers is an important marine benthic nutrient recycler and plays an essential role in marine ecosystem (Li et al. 2021b). Furthermore, numerous studies have suggested that sea cucumber may be a suitable model organism for studying the effects of environmental stresses on marine benthos due to its high sensitivities to environmental changes (Li et al. 2020, 2021b; Telahigue et al. 2020; Huo et al. 2018; Rabeh et al. 2019). Therefore, in the present study, we exposed sea cucumber to sea water (control) and Merey crude oil WAF solutions (10, 20, 40, 60, 80 and 100% nominal concentration) for 72 h, superoxide dismutase (SOD) activity, reduced glutathione (GSH) content and malondialdehyde (MDA) content in the respiratory tree of sea cucumber were detected at exposure time of 24, 48 and 72 h to evaluate the effects of crude oil on antioxidant response in sea cucumber. Our results are essential as evidence of ecological damage assessment of oil spill accidents to sea cucumber culture industry.

3.2 Materials and Methods

3.2.1 Animals

Sea cucumbers *Apostichopus japonicas* (wet weight 62.56 ± 8.74 g) were obtained from coastal aquaculture area in Dalian, China. All animals were acclimated in recirculating aquaculture systems (water temperature 16.0 ± 0.5 °C, pH 7.9 ± 0.2, salinity 33 ± 1.0 psu, dissolved oxygen 7.1 ± 0.3 mg·L^{-1}, and photoperiod 14/10 h (light/dark)) indoors with filtered and aerated seawater. During the acclimation period (7 d), sea cucumbers were fed once a day with a formulated diet.

3.2.2 Experimental Design

Merey crude oil (a heavy crude oil with 17.41° API gravity) was obtained from Dalian Petro Co., Ltd., China. The natural seawater (salinity: 32.0 ± 1.0 psu) was collected from the Xinghai Park, Dalian, China, and filtered with 0.45 μm filters. Protocols for the preparation of WAF solutions followed the method (Singer et al. 2000) described with some modifications (Li et al. 2019). WAF solutions were prepared with the seawater at an oil loading of 25 g·L^{-1} in a 10 L glass aspirator bottle (leaving approx. 20% headspace). A magnetic stirrer (ca. 100 rpm, no vortex) with a Teflon stir bar in the bottle was employed to mix sea water and crude oil for 23.5 h in the dark at 16 ± 0.5 °C, and then the mixture was settled for 30 min to separate the residual floating oil. WAF stock solutions were collected from the bottom of aspirator bottle and immediately diluted to 10, 20, 40, 60, 80, and 100% with filtered natural seawater, for the exposure tests. Total petroleum hydrocarbons (TPH) of WAF solutions were measured by the microplate ultraviolet–visible spectrophotometer according to the SAC GB 17,378.4-2007 method (SAC 2007).

After acclimation, healthy sea cucumbers were selected and randomly allocated to rectangular glass containers (15 per treatment) with a volume of 10 L WAF solution. Natural seawater (10 L) group was set as the control. During the experiment, containers were sealed to reduce hydrocarbon volatilization. Every 5 sea cucumbers selected randomly in each group were dissected to obtain respiratory trees at exposure time of 24 h, 48 h and 72 h, respectively. After washed with pre-cold 100 mM potassium phosphate (PBS) buffer (pH 7.4), all tissue samples were frozen rapidly in liquid nitrogen and stored at −20 °C and then immediately used for the subsequent biochemical analysis.

3.2.3 Determination of Biomarkers

To assess the antioxidant response to crude oil in the respiratory tree of sea cucumber, we detected SOD activity, GSH content, and MDA content, using the assay kits purchased from Nanjing Jiancheng Bioengineering Institute in China. The procedures followed the manufacturer's protocol. The crude extract homogenate of tissue sample with PBS buffer was 1:9 (*w/v*). ROS level was measured by a 2′,7′-dichlorodihydrofluorescein (DCFH-DA) method (LeBel et al. 1992) with some modifications and was described as the ratio of fluorescence intensity to protein concentration, determined with the Total Protein Assay Kit (Bradford method). SOD activity was measured based on the inhibitable reduction of a water-soluble tetrazolium salt (WST-1) method. One SOD activity unit was defined as the amount of enzyme corresponding to 50% inhibition rate of SOD. The SOD activity was expressed as $U{\cdot}mgprot^{-1}$. GSH content was evaluated with 5,5′-dithiobis-2-nitrobenzoic acid (DTNB) method (Anderson 1985) with some modifications and defined as the amount of DTNB consumed per mg of protein per min. The GSH content was expressed as $\mu mol{\cdot}gprot^{-1}$. MDA content was measured with thiobarbituric acid (TBA) method (Schmedes and Hølmer 1989) previously described and expressed as $nmol{\cdot}mgprot^{-1}$.

3.2.4 Statistical Analysis

The differences between treatments and control were determined by one-way analysis of variance (one-way ANOVA). Normality and homogeneity of each variance were checked respectively prior to the test. All results were expressed as mean ± standard deviation (SD). Differences were considered significant if $p < 0.05$. Data analysis and graphing were conducted by SigmaPlot Ver 14.0 (Systat Software, Inc., USA).

3.3 Results

Figure 3.1 presents the TPH concentration in WAF exposure solutions and the changes of biomarkers in the respiratory tree of sea cucumber with exposure time (24, 48, and 72 h) at different WAF nominal concentrations.

Figure 3.1a showed that TPH concentration of WAF exposure solutions increased with the elevated nominal concentrations and peaked at 2.32 ± 0.07 $mg{\cdot}L^{-1}$ in 100% concentration (undiluted WAF exposure solution) group. The linear fitting result manifested that the TPH concentration was strongly correlated with the nominal concentration of WAF ($R^2 = 0.968$). Figure 3.1b showed that SOD activity in the respiratory tree of sea cucumber was relatively higher at 72 h than that at 24 and 48 h. SOD activity in 40% nominal concentration group was obviously higher than that in others at the same exposure time. Figure 3.1c revealed that GSH contents

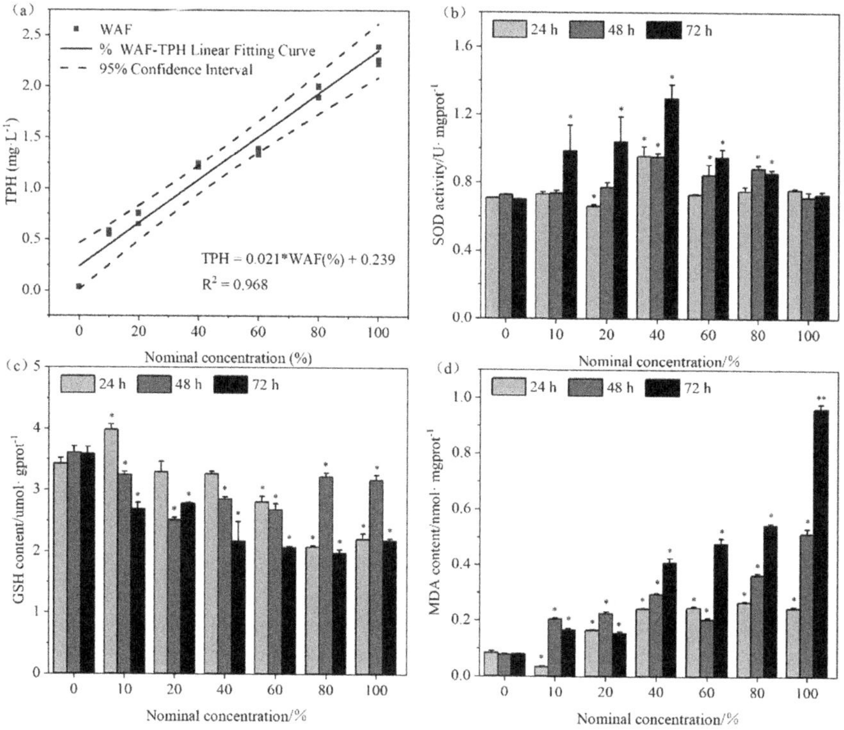

Fig. 3.1 TPH concentration (**a**) in WAF solutions and the changes of SOD activity (**b**), GSH content (**c**), and MDA content (**d**) in the respiratory tree of sea cucumber at different exposure time. * and **indicate significant differences with $p < 0.05$ or $p < 0.01$

in each nominal concentration treatment gradually decreased with continuous exposure. High nominal concentration treatments presented lower GSH contents than other groups. Figure 3.1d showed that a significant concentration-dependent increase in MDA content was observed in WAF, indicating that WAF exposure induced significant lipid peroxidation in the respiratory tree of sea cucumber.

3.4 Discussion

In the present study, sea cucumbers were exposed to different nominal concentrations of Merey crude oil WAF for 72 h to evaluate the effect of crude oil on antioxidant response in the respiratory tree of sea cucumber. Results showed that TPH concentrations represented a strong concentration-dependent relationship with nominal concentrations of WAF exposure solution ($R^2 = 0.968$). Previous study appeared to be in a similar corresponding relationship as our results (Li et al. 2019).

SOD activity in the respiratory tree was observed significant elevation from exposure time of 48 h, indicating that sea cucumber might enhance antioxidative defense to mitigate oxidative stress due to WAF exposure. The report of acute exposure (48 h) of crude oil also documented the similar pattern of SOD activity in the gill of intertidal mudskipper (*Boleophthalmus pectinirostris*) (Pan et al. 2022). Although sea cucumbers are invertebrates, we believed that these results might be implicitly related considering that the two organs both have essential physiological functions such as respiration. Besides, we found that there was an obvious reduction in GSH content in the respiratory tree of sea cucumber compared to the control with continuous exposure. High nominal concentration of WAF induced more severe GSH content decreases. It indirectly suggested that Merey crude oil WAF exposure could markedly stimulate glutathione redox system, which is evidently recognized as a ubiquitous mechanism for organism to protect cells from the toxic damage of oxidative stress and xenobiotic electrophiles (Deponte 2017; Kurutas 2016). While GSH plays a central role in intracellular antioxidant metabolic processes, such as scavenging ROS, removing oxidation products, and detoxifying exogenous compounds (Hellou et al. 2012). To further investigate the oxidative damage, we detected MDA content in sea cucumber to assess lipid peroxidation. The results showed that MDA content in Merey crude oil WAF treatments increased in a dose-dependent manner, indicating that exposure to crude oil could cause oxidative damage to the biomacromolecules of sea cucumber. Consistent with our study, other studies have reported that crude oil exposure could induce severe oxidative damage of biomacromolecules in various marine organisms, e.g., sea cucumber and fish (*Carassius auratus*) (Li et al. 2020; Wang et al. 2009).

3.5 Conclusion

Overall, we found crude oil acute exposure could stimulate antioxidative defense system (increased SOD activity and decline in GSH content) of sea cucumber and finally caused significant oxidative damage (lipid peroxidation) with time-cumulative and concentration-dependent effects. In summary, our findings provide insights into the toxic effects of crude oil acute exposure on *Apostichopus japonicas* associated with antioxidant response, and as evidence of ecological damage assessment of oil spill accidents to sea cucumber culture industry.

Acknowledgements This research was funded by the National Key Research and Development Program of China, Grant number 2018 YFD0900606, and the National Natural Science Foundation of China, Grant numbers 42076215 and 42076167.

References

Anderson ME (1985) Determination of glutathione and glutathione disulfide in biological samples. Methods Enzymol 113:548–555

Beyer J, Trannum HC, Bakke T, Hodson PV, Collier TK (2016) Environmental effects of the deepwater horizon oil spill: a review. Mar Pollut Bull 110(1):28–51

Colvin KA, Lewis C, Galloway TS (2020) Current issues confounding the rapid toxicological assessment of oil spills. Chemosphere 245:125585

Deponte M (2017) The incomplete glutathione puzzle: just guessing at numbers and figures? Antioxid Redox Signal 27(15):1130–1161

Hamel J-F, Mercier A (2008) Population status, fisheries and trade of sea cucumbers in temperate areas of the Northern Hemisphere. FAO Fisheries and Aquaculture Technical Paper, pp 257–292

Hellou J, Ross NW, Moon TW (2012) Glutathione, glutathione S-transferase, and glutathione conjugates, complementary markers of oxidative stress in aquatic biota. Environ Sci Pollut Res 19(6):2007–2023

Huo D, Sun LN, Ru XS, Zhang LB, Lin CG, Liu SL, Xin XK, Yang HS (2018) Impact of hypoxia stress on the physiological responses of sea cucumber *Apostichopus japonicus*: respiration, digestion, immunity and oxidative damage. PeerJ 6:e4651

Kurutas EB (2016) The importance of antioxidants which play the role in cellular response against oxidative/nitrosative stress: current state. Nutr J 15(1):1–22

LeBel CP, Ischiropoulos H, Bondy SC (1992) Evaluation of the probe 2′,7′-dichlorofluorescin as an indicator of reactive oxygen species formation and oxidative stress. Chem Res Toxicol 5(2):227–231

Li XS, Ding GH, Xiong YJ, Ma XR, Fan YM, Xiong DQ (2018) Toxicity of water-accommodated fractions (WAF), chemically enhanced WAF (CEWAF) of Oman crude oil and dispersant to early-life stages of zebrafish (*Danio rerio*). Bull Environ Contam Toxicol 101(3):314–319

Li XS, Xiong DQ, Ding GH, Fan YM, Ma XR, Wang CY, Xiong YJ, Jiang X (2019) Exposure to water-accommodated fractions of two different crude oils alters morphology, cardiac function and swim bladder development in early-life stages of zebrafish. Chemosphere 235:423–433

Li XS, Liao GX, Ju ZL, Wang CY, Li N, Xiong DQ, Zhang YL (2020) Antioxidant response and oxidative stress in the respiratory tree of sea cucumber (*Apostichopus japonicus*) following exposure to crude oil and chemical dispersant. J Mar Sci Eng 8(8):547

Li XS, Xiong DQ, Ju ZL, Xiong YJ, Ding GH, Liao GX (2021a) Phenotypic and transcriptomic consequences in zebrafish early-life stages following exposure to crude oil and chemical dispersant at sublethal concentrations. Sci Total Environ 763:143053

Li XS, Wang CY, Li N, Gao YL, Ju ZL, Liao GX, Xiong DQ (2021b) Combined effects of elevated temperature and crude oil pollution on oxidative stress and apoptosis in sea cucumber (*Apostichopus japonicus*, Selenka). Int J Environ Res Public Health 18(2):801

Pan G, Qiu S, Liu X, Hu X (2015) Estimating the economic damages from the Penglai 19–3 oil spill to the Yantai fisheries in the Bohai Sea of northeast China. Mar Policy 62:18–24

Pan Y, Tian L, Zhao Q, Tao Z, Yang J, Zhou Y, Cao R, Zhang G, Wu W (2022) Evaluation of the acute toxic effects of crude oil on intertidal mudskipper (*Boleophthalmus pectinirostris*) based on antioxidant enzyme activity and the integrated biomarker response. Environ Pollut 292:118341

Pasparakis C, Esbaugh AJ, Burggren W, Grosell M (2019) Impacts of Deepwater Horizon oil on fish. Comp Biochem Physiol C: Toxicol Pharmacol 224:108558

Rabeh I, Telahigue K, Bejaoui S, Hajji T, Chouba L, El Cafsi M, Soudani N (2019) Effects of mercury graded doses on redox status, metallothionein levels and genotoxicity in the intestine of sea cucumber *Holothuria forskali*. Chem Ecol 35(3):204–218

SAC. GB 17378.4-2007 (2007) The specification for marine monitoring—Part 4: Seawater analysis. In: Standardization administration of the People's Republic of China, vol GB 17378.4-2007. Standards Press of China, Beijing, pp 44–45

Schmedes A, Hølmer G (1989) A new thiobarbituric acid (TBA) method for determining free malondialdehyde (MDA) and hydroperoxides selectively as a measure of lipid peroxidation. J Am Oil Chem Soc 66(6):813–817

Singer MM, Aurand D, Bragin GE, Clark JR, Coelho GM, Sowby ML, Tjeerdema RS (2000) Standardization of the preparation and quantitation of water-accommodated fractions of petroleum for toxicity testing. Mar Pollut Bull 40(11):1007–1016

Telahigue K, Rabeh I, Bejaoui S, Hajji T, Nechi S, Chelbi E, El Cafsi M, Soudani N (2020) Mercury disrupts redox status, up-regulates metallothionein and induces genotoxicity in respiratory tree of sea cucumber (*Holothuria forskali*). Drug Chem Toxicol 43(3):287–297

Wang Y, Zhou Q, Peng S, Ma Lena Q, Niu X (2009) Toxic effects of crude-oil-contaminated soil in aquatic environment on *Carassius auratus* and their hepatic antioxidant defense system. J Environ Sci 21(5):612–617

Yuewen D, Adzigbli L (2018) Assessing the impact of oil spills on marine organisms. J Oceanogr Mar Res 6(179):472–479

Chapter 4
Robust Real-Time Updating of Real-Time Flood Forecasting System Based on Kalman Filter

Huang Zhiqiang, Liu like, Shen Kaiqi, and Zhao Chao

Abstract The updating scheme with high precision and strong robustness is one of the most important factors affecting the real-time flood forecasting system. The standard Kalman filter algorithm is often used to real-time updating, because of its timeliness and strong tracking. However, it is sensitive to outliers, a small number of outliers can cause seriously collapse. In order to withstand the destruction of outliers on updating process, a robust Kalman filter method is put forward. The robust weight function is introduced to adjust the weight of the measured data recursively. By compressing the weight of the suspicious observations and resulting in a decreased filter gain, the harmful influence of the abnormal observations on the determination of the state variables can be resisted effectively and the robustness of the updating can be achieved. The performances of the proposed method have been compared with the standard Kalman filter by both data with and without outliers. The robust method shows the robust results and the filters the impact of the abnormal observations.

Keywords Real-time flood forecasting · Outlier · Robust · Kalman filter

4.1 Introduction

Imperfections of hydrological predictions, especially real-time predictions, usually stem from input uncertainty, hydrological models structure uncertainty and parameter uncertainty (Liu and Gupta 2007; Butts et al. 2004). Thus, it is necessary to reduce uncertainty and improve accuracy employing error updating methods (Yu and Chen 2005; Bogner and Pappenberger 2011).

H. Zhiqiang
Yongningjiang Affairs Center, Huangyan District, Taizhou 318020, China

L. like · S. Kaiqi · Z. Chao (✉)
School of Environmental Science and Engineering, Xiamen University of Technology, Xiamen 361024, China
e-mail: zhaocfang@163.com

J. Baeyens et al. (eds.), *Proceedings of 2022 4th International Conference on Environment Sciences and Renewable Energy*, Environmental Science and Engineering, https://doi.org/10.1007/978-981-19-9440-1_4

Based on different updating objects, error updating methods can be divided into four groups: (1) updating inputs, (2) updating state variables, (3) updating model parameters, (4) updating model outputs according to the differences between the predicted values and the observed values. According to WMO (1992), among the four error updating methods, output updating methods are the most popularly used, as they are unconcerned in the structure of the forecasting system and can be successfully used into any forecasting models.

Kalman filter is a widely used method for real-time flood predicting system to update output, i.e., predicted discharges. The method first was proposed that it is suitable the hydrological forecasting by Hino (1970). Based on the unbiased minimum variance estimation, the filer method can obtain the optimal estimation of state variables of dynamic system for Gaussian system (Morris 1976; Gelb 1974; Todling and Cohn 1994; Wu et al. 2008). In existing literatures, Kalman filter updating process is employed to estimate the state variables and parameters of the real-time updating model, and to predict output errors. The updated outputs can then be gained by adding the predicted output errors to the forecasts (Madsen and Skotner 2005; Refsgaard 1997; Khu et al. 2001; Liu et al. 2012). Under the normal observation conditions, the method has ability to error updating and dynamic tracking (Goswani et al. 2005).

However, the assumption of the Gaussian distributions is not frequently fulfilled in most practical application because it has face to various forms of contamination of data. For example, in modern operational real-time hydrological systems, the observed data are automatically obtained by automatic collecting system. The observations often encounter abnormal situations caused by temporary instrument failures, erroneous measures, and signal disturbances (Zhao et al. 2008; Zhao and Yang 2019; Li et al. 2015). In addition, for a reservoir basin, the "observed" inflow of reservoir actually is not observed directly, but calculated by hydrologic balance equation based on the directly measured water level, outflow and level-capacity curve. In these processes, outliers, that lies outside the overall pattern of distribution (Han and Kamber 2011) and far from Gaussian, are often inevitably created. In the present of the outliers, the standard Kalman fail to perform adequately. The squared error criterion of this method is very sensitive to abnormal observations and results in low quality estimates for updating schemes (Brown and Hwang 1992).

In order to raise the robustness of the Kalman filter to outliers of non-Gaussian noise, in the past decades, a considerable number of scientific researches have focus on the development of various robust filtering approaches. Some authors have proposed several robust statistical algorithms to achieve robustness (Kitagawa 1987). However, these methods resist the effect of the outliers at the cost of their high computation complexities which are unsuitable for real-time applications. Others have made a lot of efforts to address the susceptibility of squared error criterion to outliers, yet these approaches may be quite complicated and be difficult to implement.

In this paper, an interesting robust M-estimation-based Kalman filter method is presented. It assigns less weight to outliers, so that the impact of outliers may be suppressed on state variables and parameters.

The remaining of the paper is organized as follows. In Sect. 4.2, the real-time flood forecasting correction based on standard Kalman filter is briefly introduced.

In Sect. 4.3, the robust weight function is introduced to modify the Kalman gain to make the filter robust. In Sect. 4.4, efficiency of new method is compared with the standard Kalman filter, and better performance of the proposed method is valid, followed by concluding remarks in Sect. 4.5.

4.2 Real-Time Error Updating Based on Standard Kalman Filter

The Kalman filter is a useful tool to updating measure errors and model errors of dynamic systems. It can be expressed as

$$\text{Observation function} : Z_k = H_k\theta_k + v_k \tag{4.1}$$

$$\text{State function} : \theta_k = \varphi_{k-1}\theta_{k-1} + \omega_{k-1} \tag{4.2}$$

where Z_k is measure vector which can be observed at the kth moment; H_k is observational matrix at time k; θ_k is state vector at times k and is to be estimated; φ_{k-1} is propagation matrix; ω_{k-1} and v_k, is system noise and measure noise vector, respectively; ω_{k-1} and v_k are assumed zero-mean uncorrelated Gaussian white noises satisfying $\omega_{k-1} \sim N(0, S_{k-1})$, $v_k \sim N(0, R_k)$.

Below, the symbols "∧" above variables represent estimation; the subscript "*k*/*k* − 1" and "*k*/*k*" represents the predicted and corrected estimates at time *k*, respectively.

In real-time flood updating system, the differences between the predicted discharge and the measured discharge can be used as very information for correction. Let $Z_k = e(k)$, where $e(k)$ is the discrepancy between the forecasts and observations at time k; and

$$H_k = [e(k-1), e(k-2), \ldots e(k-p)]^T$$

$$\theta_k = [\theta(1)_k, \theta(2)_k, \ldots \theta(p)_k]$$

$$\theta_{k-1} = [\theta(1)_{k-1}, \theta(2)_{k-1}, \ldots \theta(p)_{k-1}]$$

$$\varphi_k = [1, 0, \ldots 0]_{1\times p}$$

where p is the number of state variables.

Equations (4.1) and (4.2) are solved by the standard Kalman filter to give the updating estimation of the state variables, i.e., $\hat{\theta}_{k/k}$ and its covariance $P_{k/k}$, the kth Kalman filter equation are as follows:

$$\hat{\theta}_{k/k} = \arg\min\left\{\left(\hat{\theta}_{k/k} - \theta_k\right)^T \left(P_{k/k-1}\right)^{-1}\left(\hat{\theta}_{k/k} - \theta_k\right) + (Y_k - H_k\theta_k)^T (R_k)^{-1}(Y_k - H_k\theta_k)\right\} \tag{4.3}$$

In general, the method consists of two steps, including predicting step and updating step.

Predicting step:

$$\hat{\theta}_{k/k-1} = \varphi_{k-1}\hat{\theta}_{k-1/k-1} \tag{4.4}$$

$$P_{k/k-1} = \varphi_{k-1}P_{k-1/k-1}\varphi_{k-1}^T + S_{k-1} \tag{4.5}$$

Updating step:

$$\mathrm{K}_k = P_{k/k-1}H_k^T\left(H_k P_{k/k-1}H_k^T + R_k\right)^{-1} \tag{4.6}$$

$$\hat{\theta}_{k/k} = \hat{\theta}_{k/k-1} + K_k\left(Z_k - H_k\hat{\theta}_{k/k-1}\right) \tag{4.7}$$

$$P_{k/k} = P_{k/k-1} - K_k H_k P_{k/k-1} \tag{4.8}$$

where K_k is the Kalman filter gain matrix at time k.

If the Gaussian assumptions of (4.1) and (4.2) hold, the standard Kalman filter can obtain the optimal estimation of the $\hat{\theta}_{k/k}$. Then the updated errors of real-time flood forecasting system at time $k + 1$, can be predicted with Eq. (4.1) by $k + 1$ replacing k. Eventually, the updated errors are combined with the forecasts by hydrological model to achieve the real-time updating process.

However, if there is some violation to assumption about (4.1) and (4.2), e.g., that Gaussian distribution is contaminated with some other distribution, the performance of the standard Kalman filter will deteriorate considerably. The real-time updating accuracy cannot be guaranteed. A new Kalman filter algorithm insensitive to outliers will be proposed in next section.

4.3 Robust Real-Time Flood Updating

4.3.1 Robust Kalman Filter Algorithm

In this paper, the "observed" discharge consists of outlier information besides random Gaussian errors, i.e. v_k displays a mixture distribution. The mixture distribution is that a large portion of the residuals obey a normal distribution with small variance, while a small portion have an unknown distribution with much bigger variance. Hence, we can represent v_k by the mixture distribution as

$$p(v) = (1-\delta)N(v|0,\ \sigma^2) + \delta h(v) \quad 0 \le \delta < 1 \tag{4.9}$$

where $N(v|0, \sigma^2)$ denotes the zero mean normal pdf with variance σ^2, and $h(v)$ is a pdf of outliers with variance ${\sigma_h}^2 \gg \sigma^2$. δ is the probability of occurrence of the outliers.

In order to bound the influence of the outliers, redefine a robust criterion function

$$\hat{\theta}^r_{k/k} = \arg\min\left\{\left(\hat{\theta}_{k/k-1} - \theta_k\right)^T \left(P_{(k/k-1)}\right)^{-1}\left(\hat{\theta}_{k/k-1} - \theta_k\right) + \rho(Z_k - H_k\theta_k)\right\} \tag{4.10}$$

where ρ(.) is a robust score function which suppress the effect of outliers on estimation of state variables.

The weight function $\omega(x)$ is defined as $\omega(x) = \frac{\partial\rho(x)}{\partial x}/x$, The optimal estimates of parameters should be the solution of the following equations:

$$\hat{\theta}^r_{k/k} = \left(H_k^T W_k V_k^{-1} H_k\right)^{-1} H_k^T W_k V_k^{-1} Z_k \tag{4.11}$$

where $\mathrm{W}_k = \begin{bmatrix} w(M_k) & \cdots & 0 \\ \cdots & \cdots & \cdots \\ 0 & \cdots & w(M_1) \end{bmatrix}$ and $V_k = \begin{bmatrix} R(k) & \cdots & 0 \\ \cdots & \cdots & \cdots \\ 0 & \cdots & R(1) \end{bmatrix}$

$$\mathrm{P}^r_{k/k} = \left(H_k^T W_k V_k^{-1} H_k\right)^{-1}\left(H_k^T W_k V_k^{-1} H_k\right)\left(H_k^T W_k V_k^{-1} H_k\right)^{-1} = \left(H_k^T W_k V_k^{-1} H_k\right)^{-1} \tag{4.12}$$

We can obtain the desires recursive solution form for robust Kalman filter based on Eq. (4.10)

$$\hat{\theta}^r_{k/k-1} = \varphi_k \hat{\theta}^r_{k-1/k-1} \tag{4.13}$$

$$P^r_{k/k-1} = \varphi_k P^r_{k-1/k-1} \varphi_k^T + S_k \tag{4.14}$$

$$K_k^r = P^r_{k/k-1} H_k^T \left(H_k P^r_{k/k-1} H_k^T + R_k w(M_k)^{-1}\right)^{-1} \tag{4.15}$$

$$\hat{\theta}^r_{k/k} = \hat{\theta}^r_{k/k-1} + K_k^r\left(Z_k - H_k \hat{\theta}^r_{k/k-1}\right) \tag{4.16}$$

$$P^r_{k/k} = P^r_{k/k-1} - {K^r}_k H_k P^r_{k/k-1} \tag{4.17}$$

Equations (4.13)–(4.17) constitute the robust Kalman filter algorithm. It is clear that the robust filter algorithm has the similar iterative structure of the classical

Kalman filter version from Eqs. (4.4)–(4.8). The main difference is the introduction of the weight function ($w(M_k)$)). The robustness of the robust Kalman filter is focus on the weight function, which can be used to adjust weight of observation information.

The weight function is built by residual of observation vector. The following weight function has been selected

$$w(M_k) = \begin{cases} 1.0 & M_k \le k_1 \\ k_1/M_k & k_1 \le M_k \le k_2 \\ 0 & M_k > k_2 \end{cases} \tag{4.18}$$

where k_1 and k_2 are defined threshold constants. The reasonably good values of the threshold constants (Zhao et al. 2008) are $k_1 = 1.5$, $k_2 = 2.5$. The weight function will automatically assign zero weight to outliers, so that can resist the impact of outliers, a less weight to the suspicious data, and unity weight to the good data. When all data is good, the Robust Kalman filter is the same as the classical Kalman filter. In Eq. (4.18), the M_k is the judgment variable, which directly determinate the robust effect. Therefore, its selection must conform to the characteristics of real-time flood forecasting. In this paper, M_k is calculated by

$$M_k = |Q_{\text{mk}} - Q_{uk}| / \left(\frac{1}{n-m} \sqrt{\sum_{i=1}^{n} w(M_k) \cdot (Q_{\text{mi}} - Q_{ui})^2} \right) \tag{4.19}$$

where, Q_{m}, Q_u are the measured discharge and updated discharged, respectively. n is the number of the samples, m is the number of the detected outliers.

From the above the Eqs. (4.13–4.19), it can be seen that the robust filter algorithm can get the optimizing error covariance and gain matrix, and cancel the impacts of the outliers and suppress the effects of the suspicious information.

The weight function is based on residual of observations, however the residual is dependent on the weight function. The iterative algorithm is required for the robust filter, the steps at time k are shown in Fig. 4.1.

4.4 Application

4.4.1 Study Basin

Qilijie basin located in southeastern China, in semi-humid regions, is chosen. The catchment covers 14,787 km^2. The observation hourly discharge data from 1988 to 1998. The predicted time is 3 h.

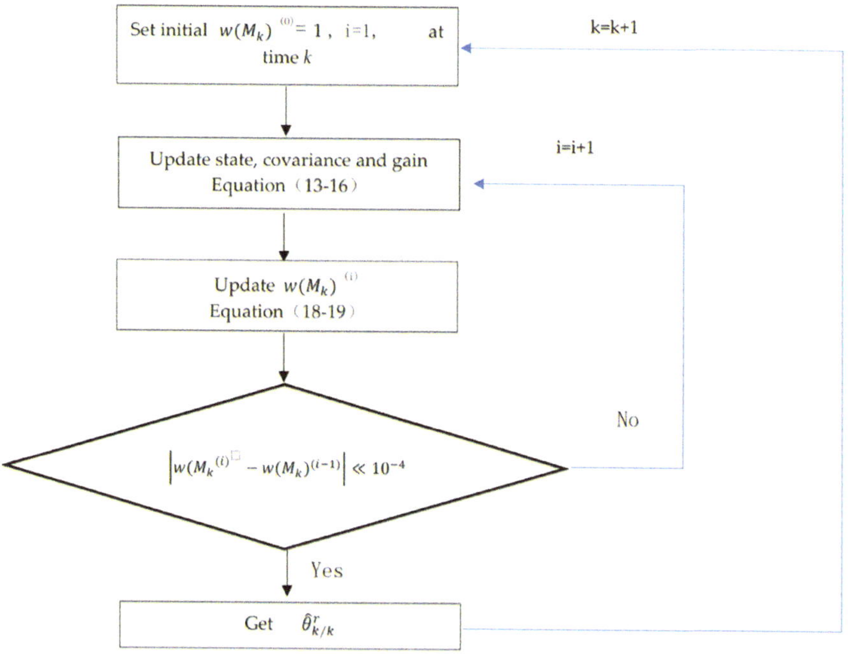

Fig. 4.1 Diagram of robust iterative algorithm

4.4.2 Hydrological Model

The Xin'anjiang model is chosen to forecast the basin outlet discharge. Xin'anjiang model was first proposed in 1960s, and has been successfully and widely employed in China. The main feature is the concept of stored-full runoff, which means that runoff is not produced until the soil moisture content of the aeration zone of reaches field capacity, and runoff equals the rainfall excess without further loss.

The basin is divided into a set of sub-basins. the outflow from each sub-basin is first simulated and then routed down the channels to the main basin outlet. The parameters of the model can be calibrated and validated based on input, output.

4.4.3 Performances with Outliers

The two Kalman filters are used to updating forecasting discharge without outlier information from 1997 to 1999. The performances were evaluated according to the following 3 indices:

(1) Nash–Sutcliffe (NS)

$$NS = 1 - \sum_{i=1}^{n} \left(Q_{mi} - Q_{fi}\right)^2 / \sum_{i=1}^{n} \left(Q_{mi} - \overline{Q}_m\right)^2 \tag{4.23}$$

(2) the relative error of flood volume

$$\Delta R = \frac{\sum_{i=1}^{n} (Q_{mi} - Q_{fi})}{\sum_{i=1}^{n} Q_{mi}} \tag{4.24}$$

(3) the efficiency of the updating method ΔR_u and NS_u

$$\Delta R = \frac{\sum_{i=1}^{n} (Q_{mi} - Q_{ui})}{\sum_{i=1}^{n} Q_{mi}}$$

$$NS_u = 1 - \sum_{i=1}^{n} (Q_{mi} - Q_{ui})^2 / \sum_{i=1}^{n} \left(Q_{mi} - \overline{Q}_m\right)^2$$

where $\overline{Q}_m$ is mean of observation discharge. Q_u is updated discharge.

NS is used to evaluate the performance of fitness between Xin'anjiang model-forecasted and observed discharge. The larger value of NS means better fitness, and the maximum value is 1.0. NS_u is used to estimate the performance of the updating scheme. The ΔR is used to assess the forecasted runoff error. The ΔR_u is used to assess the updated runoff error.

The performances of the two Kalman filters are listed in Table 4.1.

As the Table 4.1 indicates, the average NS of Xin'anjiang model forecasts without updating is 0.853. it is a satisfactory output, indicating that Xin'anjiang model is suitable for catchment. The performances of fitness of the Kalman filter are better the NS_u, which shows that both Kalman methods can improve the predicted flow according to the recent measured data without abnormal values. The ΔR of three situations are little, that displays the error flood volume is littler.

Table 4.1 Performance without outliers

Flood	No updating		Classical KF		Robust KF	
	NS	ΔR	NS_u	ΔR_u	NS_u	ΔR_u
980810	0.870	−0.029	0.942	−0.002	0.940	−0.001
980610	0.910	0.083	0.940	0.006	0.930	0.041
980402	0.890	0.025	0.950	0.004	0.953	0.003
980318	0.813	−0.021	0.954	0.001	0.952	0.002
970910	0.876	−0.096	0.970	−0.002	0.972	−0.003
970701	0.890	0.083	0.950	0.006	0.945	0
970510	0.721	−0.097	0.870	−0.002	0.880	0.024
Average	0.853	−0.007	0.939	0.002	0.939	0.009

We added the following error distribution to the above observed discharge in order to format the non-Gaussian distribution:

$$e_i = \begin{cases} (r - 0.5) \cdot \overline{Q} \cdot P & i = \text{int}(i/L)L \\ 0 & i \neq \text{int}(i/L)L \end{cases} \quad (4.27)$$

where, r is random numbers; $\overline{Q}$ is the average of observed discharge; P is a constant that controls the maximum of e, and L is the frequency of outliers. Adjusting P and L, outliers of different numbers and sizes can be generated. We generated outlier samples and ran the experiment 1000 times for (P = 1.5, L = 10), (P = 1.5, L = 15), (P = 3, L = 15), (P = 3, L = 15). The performances of the two Kalman filters on outliers are displayed in Table 4.2. It is noted that in order to compare with Table 4.1, the Q_m used in Table 4.2 to calculate the indices is the measured discharge free outlier.

It is clear that under abnormal conditions, the performance of the Kalman filter is better than that of the classical Kalman filter. Comparing Tables 4.1 and 4.2, the updating effect of classical Kalman filter is affected by the appearance of outliers, and NS_u decreases significantly. Although the change of average ΔR is not obvious, the single error value increases obviously. For example, ΔR of no. 980810 increases from −0.002 to −0.031, ΔR of no. 970910 increases from −0.002 to −0.06. At the same time, the robust Kalman filter has little change in the correction effect and is basically stable after the outliers appear.

In Table 4.2, the performance of Kalman filter of no. 970910 deteriorate obviously, the measured and updated discharge hydrographs are shown in Fig. 4.2. Q_m = the observed discharge without outliers, Q_{ukf}=the updated discharge by classical Kalman filter, Q_{urkf}=the updated discharge by robust Kalman filter.

In Fig. 4.2, the impulse noise occurred in the updated discharge by classical Kalman filter. That is because the classical Kalman filter cannot filter out artificially added outliers, and the updated discharge tracks the outlier information. However, the weight function of robust method detect the outliers and assign zero weight to

Table 4.2 Performance with outliers

Flood	No updating		Classical KF		Robust KF	
	NS	ΔR	NS_u	ΔR_u	NS_u	ΔR_u
980810	0.870	−0.029	0.912	−0.031	0.935	0.002
980610	0.910	0.083	0.939	0.007	0.932	0.021
980402	0.890	0.025	0.930	0.017	0.945	0.006
980318	0.813	−0.021	0.843	0.015	0.95	0.002
970910	0.876	−0.096	0.908	−0.060	0.965	0.003
970701	0.890	0.083	0.932	0.017	0.942	0.006
970510	0.721	−0.097	0.839	0.003	0.89	0.009
Average	0.853	−0.007	0.900	−0.004	0.937	0.005

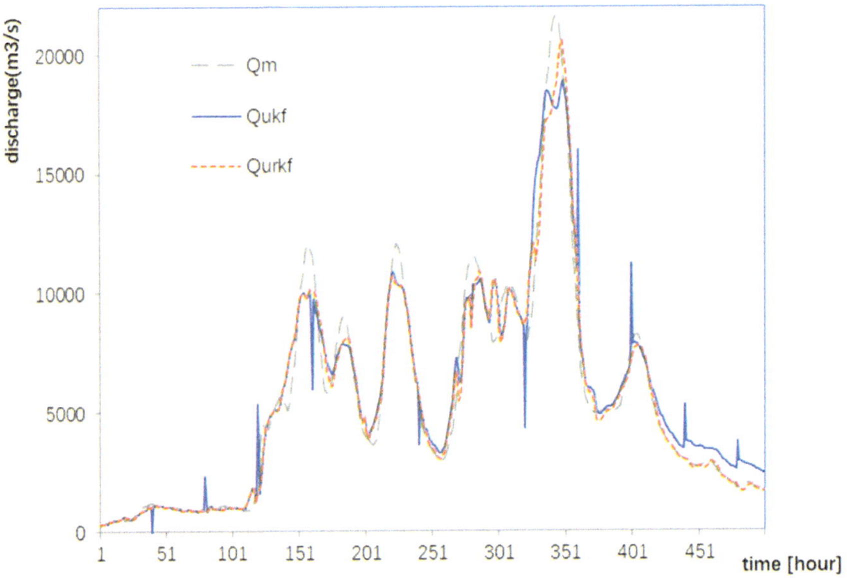

Fig. 4.2 Comparison between observed and updated discharge of no. 970910

the outlier, the discharge hydrograph got by robust Kalman filter does not appear impulse noise.

4.5 Conclusions

We propose a robust Kalman filter to update real-time flood forecasting system. To reduce the impact of outliers on accuracy of flood forecasting, the robust weight function is brought to the Kalman filter. The performance of the proposed method has been compared with the classical Kalman filter by both data with and without outliers. The robust method shows the stable updating accuracy.

In this paper, we focus our attention on the measure outliers. However, in case of real real-time flood forecasting system, the system error ω_k also may not obey gaussian distribution. We are exploring to derivative new robust Kalman filter algorithm with the simultaneously withstanding observation and state outliers.

Acknowledgements we would like to express sincere thanks to the editor and the anonymous reviewers whose comments led to great improvement of this paper.

Funding This study is funded by Natural Science Foundation of Fujian Province (2022J011232) and Scientific Research Climbing Plan of Xiamen University of Technology (XPDKT19028) and Science and technology project of Xiamen (3502Z20203063) and Innovation and start-up project of Xiamen University of Technology (YKJCX2021148).

References

Bogner K, Pappenberger F (2011) Multiscale error analysis, correction, and predictive uncertainty estimation in a flood forecasting system. Water Resour Res 47:1–24

Brown RG, Hwang PYC (1992) Introduction to random signals and applied Kalman filter: with MATLAB exercises and solutions, 2nd edn. Wiely, New York

Butts MB, Payne JT, Kristensen M, Madsen H (2004) An evaluation of the impact of model structure on hydrological modelling uncertainty for streamflow simulation. J Hydrol 298(1–4):242–266

Gelb A (1974) Applied optimal estimation. MIT Press, Cambridge Mass.

Goswani M, O'Connor KM, Bhattarai KP, Shamsedlin AY (2005) Assessing the performance of eight real-time updating models and procedures for the Brosna River. Hydrol Earth Syst Sci 9(4):394–411

Han JW, Kamber M (2011) Data mining: concepts and techniques, 3rd edn. Morgan Kaufmann Publishers, USA

Hino M (1970) Runoff forecasts by linear predictive filter. J Hydraul Div 963:681–702

Khu ST, Liong SY, Babovic V, Madsen H, Muttil N (2001) Genetic programming and its application in real-time runoff forecasting. J Am Water Resour Assoc 37(2):439–451

Kitagawa G (1987) Non-Gaussian state-space modeling of nonstationary time series. Am Stat Assoc 82(400):1032–1041

Li Q, Bao WM, Qian JL (2015) An error updating system for real-time flood forecasting based on robust procedure. KSCE J Civ Eng 19(3):796–803

Liu Y, Gupta HV (2007) Uncertainty in hydrologic modeling: toward an integrated data assimilation framework. Water Resour Res 43:1–18

Liu Y et al (2012) Advancing data assimilation in operational hydrologic forecasting: progress, challenges, and emerging opportunites. Hydrol Earth Syst Sci 16(10):3863–3887

Madsen H, Skotner C (2005) Adaptive state updating in real-time river flow forecasting—a combined filtering and error forecasting procedure. J Hydrol 308(1):302–312

Morris JM (1976) The Kalman filter: a robust estimator for some classes of linear quadratic problems. IEEE Trans Inf Theory 22:526–534

Refsgaard JC (1997) Validation and intercomparison of different updating procedures for real-time forecasting. Nordic Hydrol 28(2):65–84

Todling R, Cohn S (1994) Suboptimal schemes for atmospheric data assimilation based on the Kalman filter. Mon Weather Rev 122:2530–2557

World Meteorological Organisation (WMO) (1992) Simulated real-time intercomparison of hydrological models. Operational hydrology Rep. 1992, 38, Geneva

Wu XL, Wang CH, Chen X, Xiang XH (2008) Kalman filtering correction in real-time forecasting with hydrodynamic model. J Hydrodyn 20(3):391–397

Yu PS, Chen ST (2005) Updating real-time flood forecasting using a fuzzy rule-based model. Hydrol Sci J 50(2):265–278

Zhao C, Hong HS, Bao WM, Zhang LP (2008) Robust recursive estimation of auto-regressive updating model parameters for real-time flood forecasting. J Hydrol 349(5):376–382

Zhao C, Yang JY (2019) A robust skewed boxplot for detecting outliers in rainfall observations in real-time flood forecasting. Adv Meteorol 1795673

Chapter 5
Effect of Compost and Humus of Organic Solid Waste on the Reduction of Cadmium in the Soil and in Different Organic of Seedlings *Theobroma cacao* (CACAO) in Nursery

S. Zavala, A. Da Cruz, J. Zavala, S. Camargo, and N. Balbin

Abstract It was carried out in the facilities of the Faculty of Agronomy of the National Agrarian University of the Jungle, the objective was to evaluate the effect of compost and humus obtained from the transformation of organic solid waste in the reduction of concentration of cadmium available and total in soil and total cadmium in organs of cocoa seedlings, with a complete random design with 15 repetitions, with factorial arrangement. The soil worked is alluvial, sandy loam with 2.3 ppm of total cadmium, inoculating it with cadmium sulfate (45 days) to obtain 4.86 ppm, we proceeded to fill the bags at a proportion (20/30/40/50/60 and 70%) of compost and humus, to sow a pre-germinated seed of *Theobroma cacao L.* The results show that compost and humus statistically have the same effect on the concentration of cadmium available in the soil. Cadmium would have been adsorbed by organic matter, and would form stable complexes with humic acids, fulvic acids, humines, carboxylic acids and methoxylic acids. The application in high concentrations of organic matter would affect the growth of cocoa negatively, having obtained greater growth in treatments without cadmium inoculation and without organic matter.

Keywords Soil · Compost · Cadmium · Humus · Material

S. Zavala (✉) · A. Da Cruz · J. Zavala
Universidad Nacional Agraria de La Selva, Tingo María, Perú
e-mail: sandra.zavala@unas.edu.pe

S. Camargo · N. Balbin
Universidad Continental, Huancayo, Perú

J. Baeyens et al. (eds.), *Proceedings of 2022 4th International Conference on Environment Sciences and Renewable Energy*, Environmental Science and Engineering, https://doi.org/10.1007/978-981-19-9440-1_5

5.1 Introduction

The chemical properties of soil have a very important role in increasing or reducing the toxicity of metals, the distribution of these in soil profiles and their availability are regulated by the characteristics of the metal and its properties. Cadmium (Cd) is a heavy metal not essential for plants and very toxic to humans, whose concentration in the soil increases progressively, due to anthropic activities, such as mining, metal smelting, burning of fossil fuels, use of phosphate fertilizers, manufacture of batteries, pigments, and plastics. The absorption of cadmium by plants is an important step for the entry of this into the food chain, its absorption and subsequent bioaccumulation depend on the characteristics of the metal and the physicochemical properties of soil (Prieto et al. 2009). One of the main routes of human exposure to cadmium is the intake of contaminated food, such as chocolate, obtained from the cocoa plant. In an evaluation carried out from 2003 to 2007 by the Scientific Panel on Contaminants in the Food Chain (CONTAM), experts found that chocolate was one of the foods with a high concentration of cadmium, being the highest, so strict rules were established to minimize exposure to this metal. In research work carried out with cocoa cultivation, within the Huánuco region, as Ottos points out reports' plots with 3.65 ppm of total cadmium in the Supte area (Ottos 2018). In this way Huaynates in 2013 reports plots with total cadmium values of 3.6 ppm (Huaynates 2013). As Cárdenas expresses, in 2012 it reported cadmium values available in the soil between 1.82 and 1.63 ppm on the banks of the Huallaga and Tulumayo rivers, respectively (Cárdenas 2012). As Huamani points out in his research paper that the average available cadmium in soil is 0.53 ppm (Huamani et al. 2012). Therefore, different strategies for the reduction of cadmium bioaccumulation are currently being investigated. In the Forum "Factors associated with the bioaccumulation of cadmium in cocoa and its mitigation strategies", Ramtahal presented up to 8 possible ways to mitigate the absorption of cadmium in cocoa plants, one of these strategies being the fixation of cadmium in the soil through amendments, organic matter and/or microorganisms. Therefore, it is about taking advantage of the compost and humus obtained from organic solid waste to reduce the concentration of cadmium in the soil solution, as well as in the different organs of cocoa seedlings, such as in the root, stem, leaves and seeds. In this way, the concentration of Cadmium in the soil and in the cocoa, seedlings would be reduced, as well as taking advantage of the organic solid waste that is currently being disposed of inappropriately directly to the Huallaga River, contaminating the water, soil, and air. Therefore, the following question arises: What is the effect of compost and humus obtained from organic solid waste on the reduction of cadmium in the soil and in the different organs of cocoa seedlings?

5.2 Literature Review

5.2.1 *Soil*

Soil is composed of solid substances, water, and air. Solid substances are the residues of plants, living or dead animals and minerals that come from the disintegration and decomposition of rocks. In the water the minerals of the soil are dissolved so that the roots of the plants can take them (MOCOA 2012). Without air in the soil the roots of plants and the small animals that live in it die. Most soil experts consider the ideal soil to be made up of 50% solid material (45% mineral particles and 5% organic matter), 25% water, and 25% air (Plaster 2000).

5.2.2 *Origin of Cadmium in Soil*

One of the ways of anthropogenic incorporation of cadmium into agricultural soils is phosphate fertilization. Phosphoric rocks, which are the raw material for all phosphate fertilizers, contain levels of heavy metals that vary according to their geographical origin, but are generally higher than the average of the earth's crust. Metals remain in a significant proportion in industrial fertilizers and are subsequently applied to the soil along with phosphorus (Herrera 2010).

5.2.3 *Availability of Cadmium in the Soil*

As Sánchez expresses, he states that cadmium and most of the heavy metals incorporated into the soil can follow four different routes: They can be retained in the soil, either dissolved in the soil solution or fixed by adsorption and precipitation processes (Sánchez 2003). These processes are important since the total cadmium content in the soil gives an idea of the level of contamination, but it is the fraction of cadmium assimilable by the plant, which indicates the degree of potential toxicity of the element to living beings.

5.2.4 *Organic Fertilizers*

Biomass, is the totality of organic substances of living beings (animals and plants), can be used as a renewable raw material, as well as material energy, originates biogas and an ingenious use of biomass is the production of organic fertilizer (López 2014).

5.3 Materials and Methods

5.3.1 *Place of Study*

The present research work was carried out in the facilities of the Farm of the Faculty of Agronomy of the National Agrarian University of the Jungle (UNAS), located in the district of Rupa Rupa, Province of Leoncio Prado, department of Huánuco, between the months of November 2017 to April 2018.

5.3.2 *Equipment and Materials*

Soil flocculator of 33 samples, Polyethylene rack of 11 samples, Micropipette up to 5 ml capacity, dispenser layers of dispensing up to 25 ml, Test tubes, Tube holder rack. Paper Filter Whatman EDTA 0.05 M, pH 7, distilled water. Hydrochloric Acid 35%, Nitric Acid 65%—Reaction balloon, Graduated balloon.

5.3.3 *Calculation and Inoculation with Cadmium to the Initial Substrate*

Because, in other research works of the Huánuco region, such as Ottos where he reports plots with 3.65 ppm of total cadmium in the Supte area (Ottos 2018), as well as Huaynates who report plots with total cadmium values of 3.6 ppm (Huaynates 2013) (Table 5.1).

Table 5.1 Calculations of the amount of cadmium sulfate that was weighed to prepare a concentration of 1000 ppm of Cd

CD	3CdSO4.8H2O
3 Cd	3CdSO4.8H2O
337.2 mg Cd	769.54 mg CdSO4.8H2O
1000 mg Cd	dSO4.8H2O

5.4 Results

5.4.1 Effect of Compost and Humus of Organic Solid Waste on the Concentration of Available and Total Cadmium

In the research it was obtained as a result that the average concentration of total Cd in soil extracted according to USEPA 3050B in all samples was 2.71 ppm, and the available cadmium 1.06 ppm extracted with EDTA, 0.05 M, pH 7 (Table 5.2).

As for the levels of organic matter (%) applied, 5 groups were obtained, where the treatment with application of 40% of organic matter has the lowest average, which means that it accumulated less cadmium available and forms a single group, so statistically it is the best treatment (Table 5.3).

Table 5.2 Total cadmium and available in soil samples after extraction of plants

Treatments	n	Total (USEPA 3050B)	Available (EDTA, 0.05 M, pH 7)
T1	4	3.15	1.37
T2	4	2.98	1.04
T3	4	2.28	0.33
T4	4	2.48	0.61
T5	4	2.60	0.82
T6	4	2.84	1.05
T7	4	2.87	1.18
T8	4	2.80	0.82
T9	4	2.53	0.52
T10	4	2.56	0.81
T11	4	2.83	0.90
T12	4	2.87	1.15
T13	4	3.13	2.93
T14	4	2.00	1.37
Average		2.71	1.06

Table 5.3 Duncan's test of cadmium available in soil (ppm) at six levels (%) of organic matter

Organic matter levels (%) (B)	Half	n	Significance (*)
40	0.42	8	a
50	0.71	8	b
60	0.86	8	b c
30	0.93	8	c
70	1.10	8	d
20	1.27	8	e

Table 5.4 Concentration of cadmium in leaves, stem, and root of cocoa seedlings

Treatment	Cadmium leaves	Cadmium stem	Cadmium root
T1	2.05	1.77	1.25
T2	2.97	2.80	1.57
T3	2.12	2.65	1.30
T4	0.95	2.00	1.15
T5	0.92	1.25	0.55
T6	0.75	0.87	0.72
T7	2.07	2.42	1.20
T8	1.72	2.55	1.15
T9	1.42	2.30	0.62
T10	1.17	2.17	0.97
T11	1.02	2.07	0.65
T12	0.75	0.90	0.40
T13	2.30	2.75	1.05
T14	0.00	0.02	0.00
Average	1.45	1.89	0.90

5.4.2 *Effect of Organic Matter on the Absorption of Cadmium by Cocoa Leaves, Stems, and Root*

The average concentration of total Cd extracted dry in all samples of cocoa seedling leaves was 1.45 ppm, in stem 1.89 ppm and in root 0.90 ppm, evidencing a high accumulation of cadmium by cocoa. Table 5.4 demonstrates how these concentrations differ.

5.4.3 *Effect of Organic Matter on Cocoa Growth in Nursery*

With the application of compost, humus, and additional control (contaminated substrate (Cd), without M.O) the growth of the cocoa seedlings was statistically the same. The highest plant height was for t14, additional control (uncontaminated substrate (Cd), without M.O) with 45.09 cm, in Table 5.5.

5.4.4 *Effect of Chemical Variables (pH, M.O, CIC) on Cadmium Available in Soil*

Table 5.6 shows the Pearson correlation coefficients of available Cd and total cadmium with soil variables such as organic matter, cation exchange capacity and hydrogen potential (M.O, CIC, pH) for soil samples.

Table 5.5 Duncan of plant height of factor A

A (M.O)	Stocking	n	Significance (*)
Additional control (uncontaminated substrate (Cd), without M.O)	45.09	4	a
Additional control (contaminated substrate (Cd), without M.O	29.41	4	b
Humus	29.21	24	b
Compost	28.69	24	b

Table 5.6 Pearson correlation coefficient of the available and total Cd with the soil variables M.O, CIC, soil pH

	Cd available (Ppm)	M.O (%)	CIC (Cmol(+)/kg)	pH
Cd available	–	−0.663	−0.473	0.155
Total CD	0.492	0.220	0.350	−0.170

5.5 Discussion

The average concentration of total Cd in soil extracted according to USEPA 3050B in all samples was 2.71 ppm, and the available cadmium 1.06 ppm extracted with EDTA, 0.05 M, pH 7, where the available cadmium represents 39% with respect to the total cadmium, according to Reyes Y María in organic cocoa soils in the Dominican Republic the cadmium available was 33% (Reyes and Maria 2004). TECHNOSERVE and CITE CACAO found that the average content of available cadmium, extracted with EDTA, 0.05 M, pH 7 in soils of the San Martín Region was 0.11 ppm (TECNOSERVE and CITE-CACAO 2011). In cocoa plots Leoncio Prado Cárdenas reported on average 0.66 ppm of available cadmium (Cárdenas 2012). Huamani, in 22 plots of organic cocoa in 7-year production in Huánuco report 0.53 ppm of available cadmium (Huamani et al. 2012). Arevalo found 0.53 ppm of Cd available in Piura (Arevalo et al. 2016). In Ecuador Chavez reported 0.45 ppm of available cadmium, removable with Mehlich 3 (M3) (Chavez et al. 2016).

5.6 Conclusion

Finally, the application of compost and solid waste humus has reduced the concentration of cadmium available in the soil. Also, at higher levels they have shown greater retention (presence) of total cadmium. The application of compost and solid waste humus have reduced the absorption of cadmium by 29% (compost) and 40.79% (humus) in leaf, 31% (compost) and 25% (humus) in stem, with greater presence of cadmium in root. The organic matter presented a highly significant inverse correlation with the available cadmium and cation exchange capacity (CIC), the pH did not present a significant correlation. having obtained greater growth of the seedlings in the treatments without inoculation of cadmium and without organic matter. For the reduction or mitigation of cadmium in the soil solution and in the colloidal micelle or humic clay complex, compost and humus obtained from organic solid waste, at proportions less than 40%, should be used.

References

Arevalo E, Obando M, Zuñiga L, Arevalo C, Baligar V (2016) Heavy metals in soils of Cocoa plantations (Theobroma cacao L.) in three regions of Peru. Appl Ecol 81–89

Cárdenas A (2012) Presence of cadmium in some organic cocoa plots in the Cooperativa Agraria Industrial Naranjillo–Tingo María–Peru, Tingo Maria: thesis agronomist. Tingo María, Peru, p 113

Chavez E, Stoffella F, Baligar V (2016) Evaluation of soil amendments as a remediation alternative for cadmium contaminated soils under cacao plantations. Environ Sci Pollut Res 23:17571

Herrera T (2010) Contamination with cadmium in agricultural soils

Huamani H, Huauya M, Mansilla L, Florida N, Neira G (2012) Presence of heavy metals in cocoa cultivation (Theobroma cacao L), organic, 7th edn, pp 339–344

Huaynates J (2013) Effect of organic matter on the absorption of cadmium by the soil, in the town of Supte. In: Selva UNA (ed) Thesis Eng. RNR-mention soil and water conservation, p 111

López R (2014) Alternative program for the efficient participatory integral management and management of solid waste in the City of Tarma. Master's thesis, Universidad Nacional Mayor de San Marcos

MOCOA (2012) El suelo Propiedades físicas-químicas -Conservación, Ministerio de cultura y desarrollo rural, programa nacional de transferencia de tecnología agropecuaria pronatta, pp 1–14

Ottos E (2018) Densymmetric and chemical fractionation of organic matter associated with cadmium in alluvial and residual cocoa soil. Thesis M.Sc., Tingo Maria. In: Science in agroecology, p 111

Plaster (2000) Soil science and its management

Prieto J, González C, Roman A, Prieto F (2009) Contamination and phytotoxicity in plants by heavy metals from soils and water. In: Decima (ed) Tropical and subtropical agroecosystems, pp 29–44

Reyes E, Maria A (2004) Content of toxic heavy metals (nickel, lead, copper, cadmium and manganese) in cocoa in the province Monseñor Noul. In: Cocoa. Research results. Dominican Institute of Agricultural and Forestry Research, Santo Domingo, pp 62–73

Sánchez M (2003) Determination of heavy metals in soils of Medianadel Campo (Valladolid): extractable contents, background and reference levels. University of Valladolid, p 298

TECNOSERVE, CITE-CACAO (2011) Cadmium and lead content in cocoa production in San Martin, p 74

Chapter 6
Influence of Different Particle Sizes of Sediment Laden Flow on Erosion Rate of Tailings Dam

Jing-Yu Zhao, Jia-Ming Chang, Jia-Jia Song, and Chi-Min Shu

Abstract There are numerous safety problems of tailings pond in China, and overtopping dam break accidents occur frequently. Based on the mechanics of sediment movement, the influence of different particle sizes on the erosion rate of tailings dam in sediment laden flow was studied by means of indoor test and mathematical induction. The self-made test system for erosion characteristics and scouring process of soil slope under complex hydrodynamic conditions was used. Tailings of Nanchuan in Chongqing and Xiaoda Goose in Yunnan were used to screen out five different particle sizes (0.075, 0.106, 0.150, 0.250, and 0.425 mm) and proportioned into a sediment laden flow with a concentration of 5% for tailings erosion test. The average erosion rate was calculated and the line chart was drawn by measuring the erosion height and erosion time of tailings samples and repeating the test. The analysis drew the following conclusions: (1) With the same sediment concentration and flow velocity, the smaller the particle size was, the greater the erosion rate was. (2) The relationship curve between particle size and erosion rate showed that when the particle size changed from small to large, the gradient of erosion rate changed from large to small.

Keywords Erosion characteristics · Scour process · Sediment laden flow · Tailings dam

J.-Y. Zhao (✉) · J.-M. Chang · J.-J. Song
School of Safety Science and Engineering, Xi'an University of Science and Technology (XUST), 58, Yanta Mid. Rd., Xi'an, Shaanxi 710054, PR China
e-mail: zhaojingyu2014@126.com

J.-Y. Zhao
Shaanxi Key Laboratory of Prevention and Control of Coal Fire, XUST, 58, Yanta Mid. Rd., Xi'an, Shaanxi 710054, PR China

C.-M. Shu
Graduate School of Engineering Science and Technology, National Yunlin University of Science and Technology, Yunlin 64002, Taiwan

J. Baeyens et al. (eds.), *Proceedings of 2022 4th International Conference on Environment Sciences and Renewable Energy*, Environmental Science and Engineering, https://doi.org/10.1007/978-981-19-9440-1_6

6.1 Introduction

Tailings dam is a dam structure around the tailings pond for storing tailings and water (Yu et al. 2014). Apparently, tailings pond is a major hazard source. Once a flood overtopping dam break accident occurs, it will bring tremendous losses to human life, property, and the environmental. In recent years, with the rapid development of China's industry, the number of tailings ponds is also increasing. There are more than 10,000 tailings ponds of various sizes in China, and the tailings ponds that have not undergone environmental assessment and safety assessment account for 64.5 and 43.36% of the total respectively (Zhang and Liu 2019). There have been a large number of accidents due to the hazards of tailings pond in China. Each accident causes irreparable losses, and the safety of tailings pond cannot be underestimated. Among them, the accidents caused by overtopping dam break account for the majority.

Researches on dam failure due to overtopping mainly focus on the mechanism of failure, the overall process of dam failure, the establishment of corresponding mathematical models to simulate overtopping dams, and the factors affecting overtopping dam failure, such as the external slope ratio of the dam, the diameter of the dam body, the change of the water level in the reservoir and the increase of the infiltration line caused by continuous rainfall. Through comparative analysis, Zhao et al. (2015) summarized up the characteristics of tailings dam's body shape, material composition, principle and mechanism of dam failure, and tailings sand filling method. Deng et al. (2017) simulated the process of tailings dam failure through the model, the front part of the discharge was dominated by water flow and the tailing sand was moved by suspended load, and the back end was dominated by tailing sand and moved by bed load. Yin et al. (2010) proposed the concept of "relative permeability of tailings dam". Wang et al. (2018) observed the dam collapse process and the shape development process of the tailings pond through simulation experiments. Hu et al. (2018) analyzed the impact of using different particle size tailings to build a dam on the process of overtopping dam failure. Wu et al. (2017a) studied and discussed the characteristics of viscous tailings in the mesoscopic composition. Wu et al. (2017b) explained the influence mechanism of flocculation in tailings slurry on sedimentary characteristics with electric double layer theory. Zhang et al. (2019) studied the influence of particle size on the strength and stability of tailings dams. Wei et al. (2012) used the theory of shear stress (or traction stress) and the "steep ridge" scour model to study the mechanism of surface erosion and the mechanism of dam failure after overtopping of tailings dams. Zhang et al. (2011) concluded that the dam-breaking displacement of the tailings pond is related to the saturation degree of the dam body. The research experiment of Kong et al. (2021) showed that the sinking displacement of the dam body at the breach reached the maximum, and the particle size of the silt flow discharged from the breach gradually decreased with the increase of the distance from the breach, and the phenomenon of particle size classification changed from insignificant to obvious. Shrestha et al. (2019) concluded through experiments that the soil erosion rate and shear stress (flow) were linear, and the soil erosion resistance of the landslide dam to the water flow increases with the increase of the sediment concentration in the direction of the flow. Hancock et al. (2004) used the landscape evolution model to predict the loss of sediment on the entire landscape (ie, tons/ha/year), erosion methods (ie, slope erosion, gullies), and places where erosion may occur on the hillside. Rifai et al. (2018) studied the bank breach caused by the over-wave and the influence of floodplain tail water on the expansion and outflow of the breach. Mahdi et al. (2020) delved into the potential consequences of tailings dam failure through numerical simulation of the dam failure process of oil-bearing tailings dams. Pak et al. (2017) explored the performance of the tailings dam drainage system. Onitiri et al. (2017) studied the effects of particle size and particle loading on the stiffness and tensile strength of epoxy and polypropylene composites filled with iron tailings. Nasrin et al. (2014) concluded that the critical water level had a linear relationship with downstream slopes and riverbed slopes. Wu et al. (2021) experimental research concluded that under the action of landslides, the concentration of sand in the channel system continues to increase, which leads to a sharp increase in the discharge capacity of sandy water bodies and reduces the stable flow capacity of channel water. Research by Zhu et al. (2021) showed that under the condition of rainfall, the safety factor of tailings dam showed a continuous decline in the early stage, an increase in the mid-term decline, and a gradual decrease in the final safety decline and tended to be stable. Hancock et al. (2021) believed that collecting rainwater and draining it can effectively reduce the erosion degree of tailings dams. Chi et al. (2020) believed that the increase in tailings particle size gradually changed the damage mode of tailings to the dam body from undercut erosion to surface erosion. Chen (2020) found through research that the greater the amount of sand in the sand-bearing flow, the more fine-grained sediment transported by erosion. Research by Zhao (2019) showed that with the increase of fine particles in the water flow, the particle sedimentation method will also change from single particle "independent sedimentation" to flocculated "integral sedimentation". Wu (2017) established the relationship between particle size and deposition distance.

Although the research on the gradual failure process of tailings dam overtop has been mature on a macro level, there are still few studies considering the effect of different particle sizes on the erosion rate of tailings dams in sand-laden water flow. From the theory of sediment dynamics, it is found that the factors of sediment erosion also have the effect of particle size in the sand-bearing water flow. Therefore, it is of practical significance to analyze the influence of particle size changes on the erosion degree of tailings dams, which is of practical significance for in-depth research on the erosion mechanism of tailings dams.

6.2 Experimental Test

6.2.1 Sample Preparation

To prevent the tailings from being deposited on the bottom of the blending tank, which will affect the uneven concentration of sandy water flow. It was necessary to fully mix the particles selected from a red mud tailings in Nanchuan, Chongqing and Xiaoda goose tailings in Yunnan. Firstly, the existing tailings were dried, then crushed with a rock crusher, and then sieved. The 5 types of screens used this time were 200 mesh screens (0.075 mm), 140 mesh screens (0.106 mm), 100 mesh screens (0.150 mm), 60 mesh screens (0.250 mm) and a 40-mesh screen (0.425 mm). The screened tailings are shown in Fig. 6.1.

After the tailings sand screening was completed, the sand-containing water stream was formulated to a fixed concentration of 5%. First, added 10 kg of tailing sand and 190 L of clean water to the 1.2 * 0.6 * 0.5 m uncovered rectangular parallelepiped slurry concentration adjustment pool made of glass material. Then turned on the mixer in the pool. The tailings sand and the clean water were fully stirred, and the slurry concentration in the pool can be kept relatively constant.

After the sandy water flow was prepared, the tailings sample will be produced. 12 h in advance, mixed the dry full-size tailings with a moisture content of 15% (Jing 2011). First used an electronic scale to take 2 kg of full-size tailings, and then measure 0.3 kg of water, and used a watering can to evenly spray, proportion, and stir. Then covered it with a film. Used an electronic scale to weigh 349 g of tailings with 15% moisture content. Made 2 sets of samples. The length was 15 cm. The bulk density was 1.85 g/cm^3 (the red mud particle size was 0.088–0.25 mm, compared with Appendix B in the "Technical Regulations for Safety of Tailings Pond" (AQ 2006–2005)). The inner diameter of the sample tube was 4 cm. Added the sample to the sample tube in 5 times. The sample preparation used a geotechnical compaction hammer. After each compaction, a laser rangefinder was used to measure to ensure that the height after each compaction was 3 cm, which met the requirements of a density of 1.85 * 103 kg/m^3 (GB/T50123 2019). Finally, pushed the sample out to be flush with the port of the sample tube, and flattened it with a knife to ensure that the contact surface of the sample and the water flow was flat.

6.2.2 Erosion Test

Test Equipment. The test device used this time was a self-made test and test system for soil slope erosion characteristics and erosion process under complex hydrodynamic conditions. The device included five parts: A mud pump and a reservoir, a slurry concentration adjustment device, an erosion measurement device, a water flow rate control device, and a sample push device. The device used a solenoid valve to control the water flow rate, and used a transparent acrylic horizontal tube for easy observation. It can clearly recorded the tailings particle starting and particle movement through a high-speed camera, and measured and recorded the current water flow velocity with a sound wave flow meter. The schematic diagram of the device is shown in Fig. 6.2.

Test Setup. First, screen a red mud tailings from Nanchuan, Chongqing and Xiaoda goose tailings from, Yunnan. According to the requirements of "Geotechnical Test Method Standard (GB/T50123–2019)", five kinds of tailing sands with different particle diameters were screened out with specific particle sizes of 0.075, 0.106,

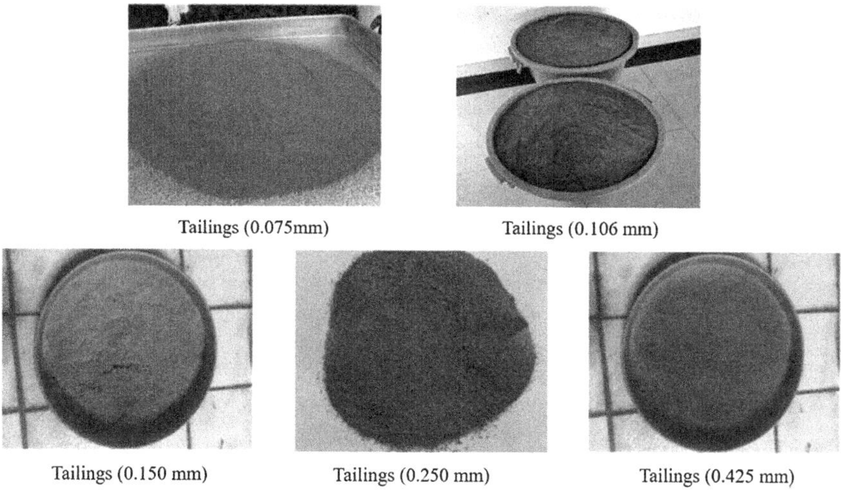

Fig. 6.1 Five particle sizes of tailings

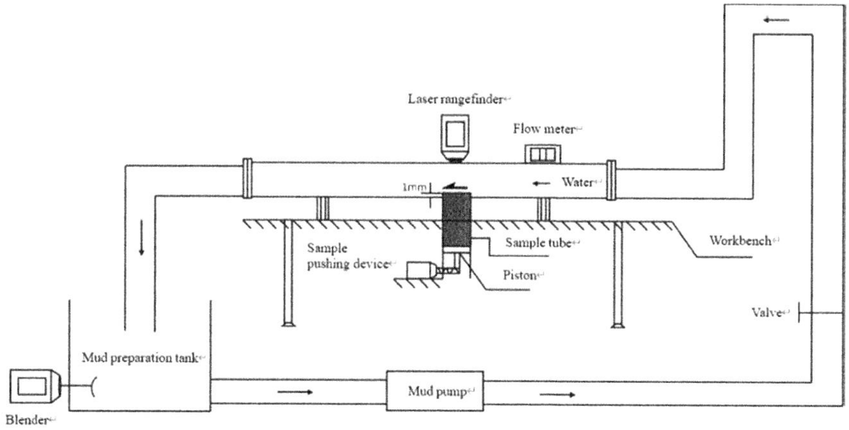

Fig. 6.2 Schematic diagram of the test device

Table 6.1 Test condition data table

Number	Concentration (%)	Velocity (m/s)	Size (mm)
1	5	0.5	0.075
2	5	0.5	0.106
3	5	0.5	0.150
4	5	0.5	0.250
5	5	0.5	0.425

0.150, 0.250 and 0.425 mm. Second, to stir the sandy water flow evenly and avoid the tailings from depositing at the bottom, it was formulated into a sandy water flow with a concentration of 5%. Then, 3 sets of repeated experiments were performed for each particle of different size using a self-made test system to reduce errors. A total of 15 groups of experiments were carried out. Finally, record the data and calculate the results. The test conditions are shown in Table 6.1.

Observation Indicators and Data Processing. The erosion measurement device consisted of a laser rangefinder and a camera. The principle of calculating the erosion rate (Fig. 6.3) was that when measuring the erosion amount of the sample, first used a laser rangefinder at the upper end of the horizontal tube to measure the height h_1 of the sample pushed into the horizontal tube. After the experiment, the laser rangefinder was used to measure the maximum erosion height h_2 of the eroded sample. The calculation formula of erosion rate z was shown in Eq. (6.1).

$$z = (h_2 - h_1)/t \tag{6.1}$$

In the formula, h_1 was the initial reading of the sample pushed into the horizontal tube, mm; h_2 was the deepest reading of the sample erosion, mm; t was the erosion time, s.

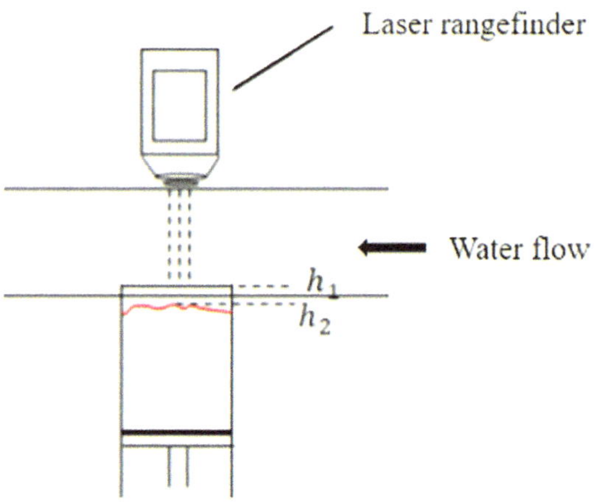

Fig. 6.3 Schematic diagram of erosion rate calculation

6.3 Results and Analysis

The following was the erosion rate diagram of different particle sizes (Figs. 6.4, 6.5, 6.6, 6.7 and 6.8), the results can clearly seen the erosion rate changes.

From one to five sets of erosion rate diagrams, it could be seen that, as a whole, the erosion rate gradually decreases (as the particle size increases). From a mechanical point of view, the condition for the initiation of sediment was that the moment of the water flow acting on the particles was greater than the moment of the underwater gravity of the particles and the cohesive force between the particles. The presence of sediment changed the force of the water flow on the particles. Gas, water, and most liquids belong to Newtonian bodies (fluids that comply with Newton's law

$$\tau = \mu d_u / d_y, \tag{6.2}$$

where μ was called the viscosity coefficient of the fluid).

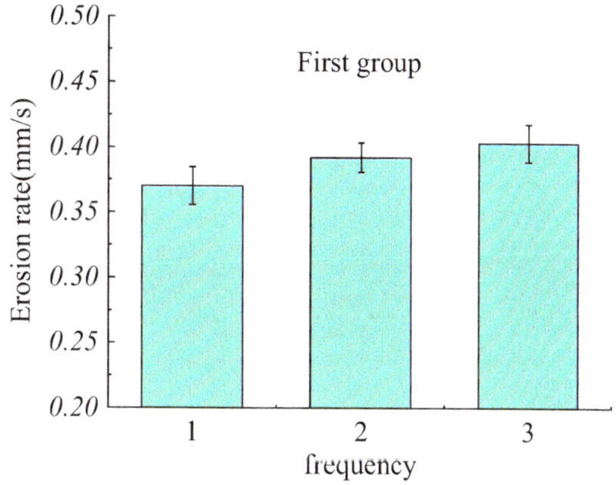

Fig. 6.4 The first group of erosion rate

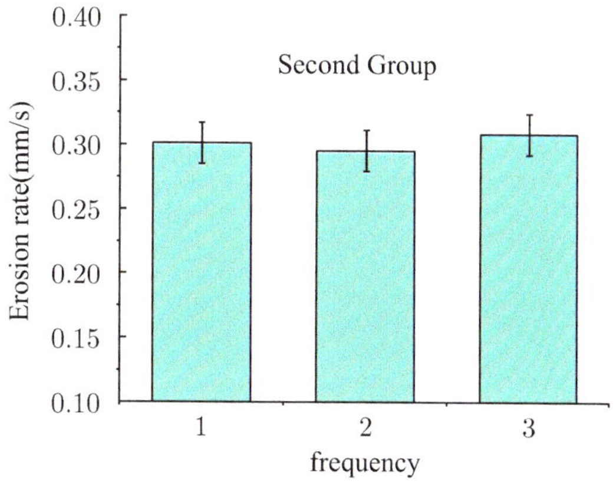

Fig. 6.5 The second group of erosion rate

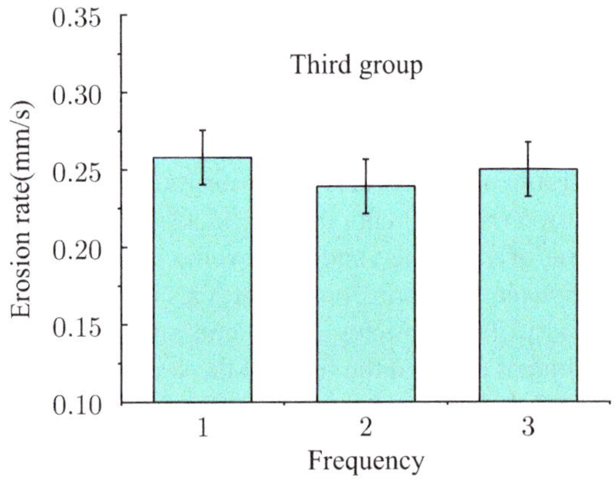

Fig. 6.6 The third group of erosion rate

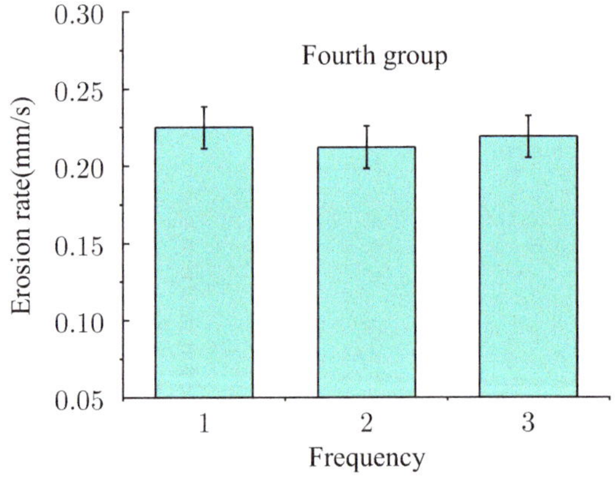

Fig. 6.7 The fourth group of erosion rate

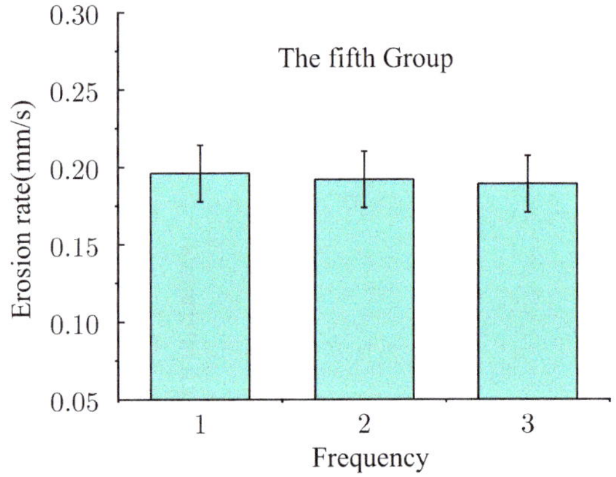

Fig. 6.8 The fifth group of erosion rate

When the water flow contains sediment, because of the deformation of the streamline, the rotation of asymmetric particles, and the flocculation of sediment increased the resistance of the water flow and increased the viscosity coefficient. As the viscosity coefficient μ increased, the shear force τ of the water flow increased, and the water flow conditions exceed the starting conditions of the sediment particles when the water was flushed. More and larger sand particles start to sport under the action of the shear force of the water flow. The moving sediment carried in the sand-bearing water flow would also collide with the stationary sediment on the bed surface to generate energy exchange, making it start to become bed load or suspended load movement. Charlotte and Carl (2004) studied the law of riverbed erosion caused by the energy exchange between the sediment particles in the water flow and the sediment particles on the bed surface during the scouring process. It was proved that increasing the sand content in the water flow under certain conditions could increase the scour intensity of the river bed. It is generally believed that the presence of suspended mass will restrain the turbulence. Yalin (1972) obtained the conclusion that the suspended mass in the water flow reduces the turbulence scale through experiments. Compared with clear water, the turbulent scale of sandy water flow was reduced, the turbulence intensity was weakened, the average size of vortex was reduced, the exchange of large and small vortices was weakened, and the exchange between suspended load and bed load was diminished.

From the viscous properties of water flow, it is known that the increase of fine particle content leads to the increase of water flow viscosity and viscosity coefficient, which in turn affects the water flow shear stress acting on the bed sand. The results are shown in Fig. 6.9.

The relationship between particle size and erosion rate was that when the particle size changed from small to large, the rate of change of erosion rate changes from large to small. In addition, when the particle size in the sand-containing water stream was smaller, the erosion rate of the tailings dam would increase instead. Conversely, when the particle size in the sand-bearing water flow became larger, the erosion rate of the tailings dam became smaller. There were two main reasons for the analysis.

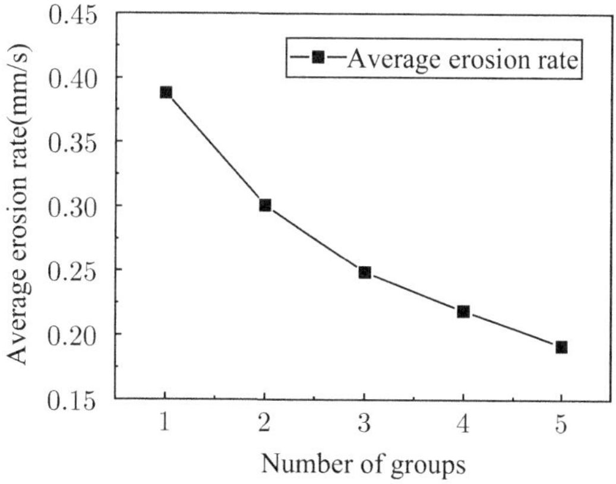

Fig. 6.9 Map of average erosion rate

Coarse sediment particles were more affected by gravity. More particles were moving in bedding state in sandy water flow. The finer the particles, the weaker the effect of gravity and the increase of the viscosity of the water flow, the slower the sedimentation speed and the more the particles moved in the suspended load state. The more the bed load, the more particles were concentrated in motion near the bottom layer. Compared with the fine particle water flow dominated by suspended load, the bottom layer had a larger density gradient and a larger flow velocity gradient. The flow velocity of the water flow on the bottom bed sand was reduced, and the drag force on the bed sand is smaller.

Adding fine particles to the water stream was more viscous than adding coarse particles. When the sand content was not large enough, the water flow was still Newtonian, and Newton's law was obeyed. When the water flow velocity remained constant, the viscosity of the water flow increased, the viscosity coefficient increased, and the shear stress between the water flows became larger, so the erosion rate of the fine-grained sand-bearing water flow was larger than that of the coarse-grained sand-bearing water flow.

6.4 Conclusion

The safety problems of tailings ponds caused by overflowing dam failure were extremely serious. In this paper, a self-made test system for soil slope erosion characteristics and scour process under complex hydrodynamic conditions was used to conduct indoor erosion tests of tailings samples. This paper started from the experiment of measuring the effect of particle size in sand-laden water flow on the erosion rate of tailings dams, and analyzes the influence of particle size changes in water flow on the erosion rate of tailings dams. It wass of practical significance to study the erosion mechanism of tailings dam overflow, and the following conclusions were drawn.

According to theoretical derivation and experimental proofs, under the same sand concentration and the same flow velocity, when the particle size is the smallest, the erosion rate is greater.

The relationship between particle size and erosion rate is that when the particle size changes from small to large, the rate of change of erosion rate changes from large to small. And when the particle size in the sand-contained water flow is smaller, the erosion rate of the tailing dam will be greater; on the contrary, when the particle size in the sand-contained water flow is larger, the erosion rate of the tailing dam will be smaller.

References

Chen Z (2020) Study on runoff and sediment and hydrodynamics characteristics in the scour process of muddy water on slope of loess engineering accumulation. D Northwest A & F University

Deng Z, Chen SS, Zhong QM (2017) Mathematical model for breach of tailings dam due to overtopping and its application. Chin J Geotech Eng 39(5):932–938

GB/T50123 (2019) Standard for geotechnical testing method. S Ministry of Housing and Urban-Rural Development of the People 's Republic of China

Gordon GD, Zhou M, Shrestha MS, Song D, Choi CE, Cui KFE, Peng M, Shi Z, Zhu X, Chen H (2019) Experimental investigation on the longitudinal evolution of landslide dam breaching and outburst floods. J Geomorphol 334:29–43

Hancock GR (2004) The use of landscape evolution models in mining rehabilitation design. J Environ Geol 46(5):561–573

Hancock GR (2021) A method for assessing the long-term integrity of tailings dams. Sci Total Environ 779:146083

Hu H (2018) Study on influence of dam constructed by tailing of different sizes on dam break due to flooding. Urbanism Architect 29:116–117

Jing XF (2011) Research on sediment flow characteristics and disaster protection of tailings dam burst. D Chongqing University

Kong XY, Sui SG, Wang GJ, Zhou HM, Zhang C, Han YB, Xu HHJ (2021) Model test study on tailings dam break under the condition of reservoir water level fluctuation. Nonferrous Met Eng 11(06):101–108

Mahdi A, Shakibaeinia A, Dibike YB (2020) Numerical modelling of oil-sands tailings dam breach runout and overland flow. J Sci Total Environ 703:134568

Nasrin J, Mahdi TF (2014) Experimental investigation into rockfill dam failure initiation by overtopping. J Nat Hazard 74(2):623–637

Onitiri MA, Akinlabi ET (2017) Effects of particle size and particle loading on the tensile properties of iron-ore-tailing-filled epoxy and polypropylene composites. J Mech Compos Mater 52(6):817–828

Pak A, Nabipour M (2017) Numerical study of the effects of drainage systems on saturated/unsaturated seepage and stability of tailings dams. J. Mine Water Environ 36(3):341–355

Rifai I, Abderrezzak K, Erpicum S, Archambeau P, Violeau D, Pirotton M, Dewals B (2018) Floodplain backwater effect on overtopping induced fluvial dike failure. Water Resour Res 54(11):9060–9073

Thompson CEL, Amogs CL (2004) Effect of sand movement on a cohesive substrate. J Hydraul Eng 130(11):1123–1125

Wang GJ, Tang YJ, Du C, Kong XY (2018) Experimental study on dam break under water level variation in reservoir. J Sediment Res 43(04):67–73

Wei Y, Xu KL (2012) Research on mechanism and process of tailing dam break due to overtopping. J Metal Mine (04):131–135

Wu SW (2017) Effect of particle size on physico-mechanical properties of tailings and dam stability study. D Chongqing University

Wu SW, Yang CH, Hu XM, Jing XF (2017a) Research on correlation of tailings particle properties and compression consolidation properties. J Huazhong Univ Sci Tech (Natural Science Edition) 45(11):121–126

Wu SW, Yang CH, Zhang C, Mao HJ, Jing XF (2017b) Indoor scale-down test of tailings. J Chin J Eng 39(10):1485–1492

Wu HM, Li C, Duan GF (2021) Study on the flow stability of sandy water in gully system under the influence of landslide. J Catastrophol 36(3):60–63

Yalin MS (1972) Mechanics of sediment transport, 1st edn. Pergamon Press, London

Yao C, Wu LG, Yang JH, Xiao LX, Liu XF, Jiang QH, Zhou CB (2020) Influences of tailings particle size on overtopping tailings dam failures. J Mine Water Environ 40(1):174–188

Yin GZ, Jing XF, Wei ZA, Li XS (2010) Study of model test of seepage characteristics and field measurement of coarse and fine tailings dam. Chin J Rock Mech Eng 29(S2):3710–3718

Yu GM, Song CW, Pan YZ, Li L, Li R, Lu SB (2014) Review of new progress in tailing dam safety in foreign research and current state with development trent in China. Chin J Rock Mech Eng 33(S1):3238–3248

Zhang JR, Liu JL (2019) The statistics and causes of dam break and leakage in Chinese tailings pond. China Molybdenum Ind 43(04):10–14

Zhang XK, Sun EJ, LI ZX (2011) Experimental study on evolution law of tailings dam flood overtopping. China Saf Sci J 21(07):118–124

Zhang C, Ma CK, Yang CH, Chen QL, Pan ZK (2019) Effects of particle diameter on shear strength of tailings and stability of tailings dams. Chin J Geotech Eng 41(S1):145–148

Zhao HG (2019) Engineering characteristics of fine-grained tailings and research on dam body stability. D Kunming University of S & T

Zhao TL, Chen SS, Zhong QM (2015) Advances in studies of tailing dam break mechanism and process. Hydro-Sci Eng (1):105–111

Zhu M, She TT (2021) Numerical simulation of dam seepage characteristics of tailings reservoir under rainfall conditions. J Water Conserv Sci Technol Econ 27(06):67–71+78

Part II
Waste Management, Waste Utilization and Sustainable Development

Chapter 7
Properties of Porous Concrete Using Expanded Polystyrene (EPS) Foam Waste as Cement Binder Admixture

C. Boonpeng, T. Suwan, W. Liu, C. Hansapinyo, and B. Paphawasit

Abstract Porous Concrete is becoming more popular and widespread as it can drain the water from the top surface to the underneath effectively. It is then applied to the road or pavement surface construction. However, some typical properties of cement-based porous concrete are limited, e.g., low strength, poor durability, or scaling-off of pavement. Conventional polymer admixtures, such as unsaturated polyester and epoxy resins, are thus used to reduce those problems. However, the conventional polymer admixture is quite expensive and less environmentally friendly. This study focuses on utilizing the Expanded polystyrene (EPS) foam wastes as an admixture to improve cement-based porous concrete properties. The results show that 10% addition of EPS foam waste admixture provided better compressive strength than that of typical porous concrete with no admixture from 12.02 to 17.89 MPa. This approach of recycling EPS foam waste could be one of the alternative admixtures for the cement and concrete sector. The outcomes could not only improve the specific property for the cement-based porous concrete as construction materials but also promote the recycling of EPS foam wastes and allow communities to be greener and cleaner.

Keywords Admixture · Cement binder · Expanded Polystyrene · Porous concrete

---

C. Boonpeng · T. Suwan (✉) · C. Hansapinyo
Department of Civil Engineering, Faculty of Engineering, Chiang Mai University, Chiang Mai, Thailand
e-mail: teewara.s@cmu.ac.th

T. Suwan
Center of Excellence in Natural Disaster Management, Chiang Mai University, Chiang Mai, Thailand

W. Liu
College of Transportation and Civil Engineering, Fujian Agriculture and Forestry University, Fuzhou, P. R. China

B. Paphawasit
College of Arts, Media and Technology, Chiang Mai University, Chiang Mai, Thailand

J. Baeyens et al. (eds.), *Proceedings of 2022 4th International Conference on Environment Sciences and Renewable Energy*, Environmental Science and Engineering, https://doi.org/10.1007/978-981-19-9440-1_7

7.1 Introduction

In the construction sector, Concrete is one of the most widely used materials. Concrete is typically made of Ordinary Portland cement (OPC), aggregates (Sand and Gravel), water, and admixture. A suitable proportion of those OPC-water-admixture (as a cementitious binder) and sand-gravel (as aggregates) could provide a terrific dense and compact cement matrix with fewer voids and entrapped air bubbles, providing very high strength properties. However, there is another type of concrete that has no fine aggregate (Sand) in the mixture, known as 'Porous concrete' (Toghroli et al. 2018). Porous concrete is a particular type of concrete that consists of cement binder and coarse aggregate (Gravel). The gravels are entirely coated with cement slurry binder and tightly adjoined to the surrounding gravels like caramelized popcorns. After the hardening process, the huge scattered air voids are left inside the porous concrete due to the lack of sand as a filler. With the presence of many voids and gaps, the unit weight and mechanical strength are significantly decreased. Nevertheless, this occurrence leads to a porous property, which allows air and water to flow through the concrete efficiently. The porous concrete is then applied for specific purposes, e.g., drainage pavement block, media for wastewater treatment, land or surface covering, or breathable plant stone (Toghroli et al. 2018; Matar and Barhoun 2020). In fact, with no sand as a filler, the strength of the porous concrete directly depends on the ability of the cement binder to hold all gravels together in the high tensile stress and shear force conditions. To improve more strength and durability, some polymer compounds as admixtures are thus applied. However, conventional polymer binders or admixtures such as unsaturated polyester and epoxy resins are quite expensive and provide low crack propagation resistance with brittleness (Matar and Barhoun 2020; Ariffin et al. 2018).

Expanded Polystyrene (EPS) foam is one of the plastic types which contains over 98% of air voids and only 2% of the plastic. It is widely used as food containers, item packaging, insulated construction material and many others, with a dramatic increase in consumed quantity since the past several decades. It is reported that over 64 million pieces of plastic and foam wastes have been used, and most of them were disposed of inappropriately (Fig. 7.1) (GISTDA 2019 The 5-years report of foam wastes 2019). The EPS foam can be deformed and returned to its primary polystyrene plastic glue (foam glue) when dissolved in some solvents, e.g., acetone and toluene. It has been reported that foam glue can also be used as an admixture to improve some hardened properties of cement-based materials (Ariffin et al. 2018; Eskander et al. 2021).

To be a part of an environmentally friendly scheme, the study on utilizing EPS foam wastes as a polymeric foam admixture has therefore been carried out. In some manners, the utilization of recycled aggregates, e.g., from building demolition, can also be used as concrete constituents (Matar and Barhoun 2020). The main findings could be one of the potential approaches for greener and cleaner construction materials like those of alternative cementitious materials alkaline activated cement (AAC),

(a) EPS foam from the food container (b) Inappropriate EPS foam disposal

Fig. 7.1 EPS foam wastes

or geopolymer cement (Maichin et al. 2020; Jitsangiam et al. 2021; Wattanachai and Suwan 2017; Bualuang et al. 2021).

The main aim of this study is to investigate the properties of porous concrete, which contains EPS foam glue as an admixture. Significant factors were focused on viz. the water-to-cement (w/c) ratio, acetone-to-toluene (A:T) ratio, and coarse aggregate gradation. The compressive strength and the porosity of the final hardened porous concrete were examined. The utilization of EPS foam waste as an admixture for cementitious binder may not only be an approach to properly eliminate wastes but also be an alternative approach to produce a value-added product and reduce the environmental impact.

7.2 Materials

The list of materials in the experiment consisted of EPS foam wastes, solvents (acetone and toluene), and local coarse aggregates (Fig. 7.2). Portland Cement (OPC) was SCG Type 1, which complied with ASTM C150 and Thai Industrial Standard (TIS) 15–1. This dark-grey powder had a specific gravity of 3.15. Its chemical compositions by X-ray fluorescence (XRF) analysis are as shown in Table 7.1. Two potential sizes of coarse aggregate, 3/8" and 1/2", were used in the test. The average bulk specific gravity of them was 2.03, while the average percent of absorption was 1.22%.

The recycling EPS foam wastes were collected from a generous donation in the community. Its dry density was in the range of approximately 15 to 500 kg/m^3, depending on the type of its production. The EPS foam waste was shredded into small pieces before dissolving in the solvent. Laboratory-grade acetone (A) was purchased from Union Science Co., Ltd. with a specific gravity of 0.79. Laboratory–grade toluene (T) was also purchased from Union Science Co., Ltd, Thailand. It was a clear liquid under ambient temperatures with a specific gravity of 0.87.

(a) Foam wastes (b) Acetone and Toluene (c) 3/8" gravel (d) 1/2" gravel

Fig. 7.2 Constituents of EPS foam admixture for porous concrete production

Table 7.1 Chemical composition of OPC by XRF analysis

Materials	Al_2O_3	SiO_2	SO_3	K_2O	CaO	TiO_2	Fe_2O_3
OPC	2.46	11.93	4.89	1.03	74.96	0.41	4.06

7.3 Mixture Designations and Analytical Methods

A liquid solvent (L) was a combination of acetone and toluene, presenting as an acetone-to-toluene ratio (A:T). EPS foam waste (F) was added into the prepared solvent with a specific liquid solvent-to-EPS foam (L:F) ratio by weight. All constituents were well-mixed together to form an EPS admixture. The coarse aggregate, a combination of 1/2" and 3/8" in sizes (1/2":3/8"), was then mixed with the cement paste in various water-to-cement (w/c) ratios. The ratio of cement paste-to-aggregate was 0.36. The EPS admixture was finally added into the fresh porous concrete mixes for 10% w/w of the cement paste. The uniform mixtures were placed into the prepared molds and kept at room temperature until reaching their testing age.

7.3.1 Mixture Designations for the Different Coarse Aggregate Sizes

Preliminary testing for early strength test (7 days age) was set to investigate the effects of coarse aggregate sizing. The variable ratios of the 1/2": 3/8" aggregates were 30:70 and 50:50, while the various w/c ratios of 0.25, 0.30, and 0.35 were applied with 10% EPS admixture. The L:F ratio was 1.5:1 and A:T ratio was 1:1. The details of the mixture and proportion for the test are as presented in Table 7.2.

Table 7.2 Details of mixture and proportion for the different coarse aggregate sizes

L:F	A:T	EPS admx	w/c ratio	1/2": 3/8"	Evaluation
1:1	1:1	10%	0.25	30:70	Early compressive strength at 7 days and Porosity
				50:50	
			0.30	30:70	
				50:50	
			0.35	30:70	
				50:50	

Table 7.3 Details of mixture and proportion for different acetone-to-toluene ratios

1/2": 3/8"	L:F	EPS admx	w/c ratio	A:T	Evaluation
50:50	1:1	10%	0.25	1:1	Compressive strength at 28 days and Porosity
				3:2	
			0.30	1:1	
				3:2	
			0.35	1:1	
				3:2	

7.3.2 Mixture Designations for the Different Acetone-To-Toluene Ratios

The effects of acetone-to-toluene ratio on the compressive strength and porosity were investigated. The variable ratios of the A:T were 1:1 and 3:2, while the various w/c ratios of 0.25, 0.30, and 0.35 were applied with 10% EPS admixture. The 1/2": 3/8" and L:F ratios were 50:50 and 1:1, respectively. See Table 7.3 for details.

7.3.3 Analytical Techniques

A compressive strength test was carried out with a 40 × 40 × 160 mm. prism specimen following EN 196–1 standard in order to examine the interfacial bonding strength of coated binder on the surface of coarse aggregates. The specimens were de-molded 24 h after casting, then wrapped with plastic film and left in the ambient conditions until reaching their ages of testing. The strength test was operated by a 250 kN Control Brand universal testing machine (UTM). The porosity of all specimens was calculated before the crushing test. The mixing process is as presented in Fig. 7.3.

(a) EPS foam admixture (b) Mixing of fresh porous concrete

Fig. 7.3 Specimen preparation process

7.4 Results and Discussions

7.4.1 Effect of the Coarse Aggregate Size on the Early Strength and the Porosity

A preliminary result for early strength of 7 days age was carried out to determine the effects of various 1/2": 3/8" aggregate sizing ratios viz. 30:70 and 50:50, respectively. The early strength and the porosity of the porous concrete are shown in Fig. 7.4. It can be seen that the early compressive strength of porous concrete with an aggregate sizing of 50:50 achieved higher strength than that of 30:70 for all w/c ratios. A more quantity of smaller aggregate size of 3/8" in the 30:70 ratio obtained less strength by the high cohesive binder, leading to an incomplete binder coating of tiny 3/8" gravels. Scattered air voids were left and increased their porosity at around 33 to 35%. The aggregate size ratio of 50:50 had larger gaps than the 30:70, allowing that viscous binder to fill in effortlessly. This occurrence led to a decrease in porosity from around 15% to 22%. The w/c ratio of 0.30 achieved optimal flowability and consistency for the fresh porous concrete. The w/c ratio of 0.25 was too dry, while the 0.35 was too wet. However, the 50:50 aggregate size with the 0.30 w/c ratio has passed the limitation of control porous concrete with no admixture of 4.48 MPa in compressive strength and 22.07% of porosity.

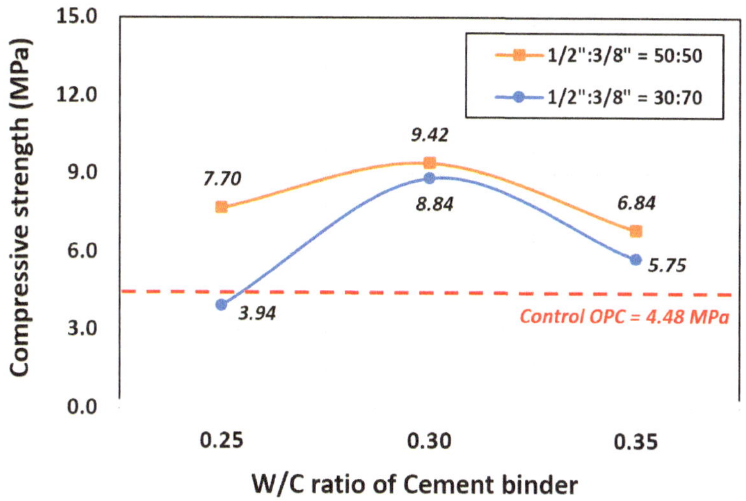

(a) Early strength of porous concrete

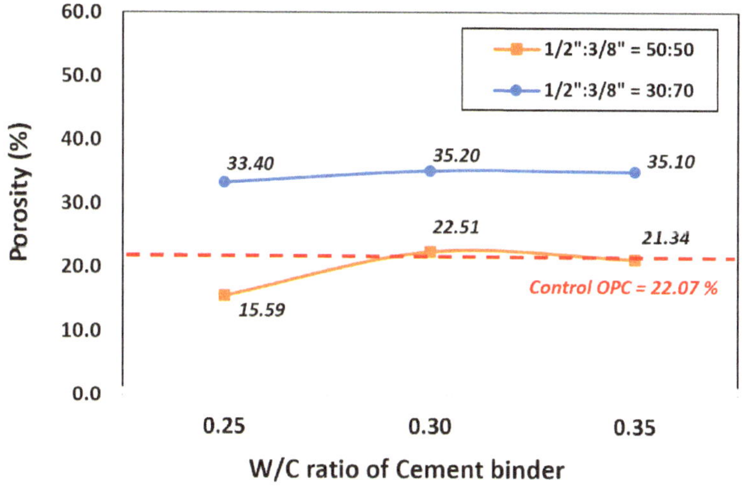

(b) Porosity of porous concrete

Fig. 7.4 Properties of porous concrete with different aggregate sizing at 7 days of age

7.4.2 Effect of Acetone-To-Toluene Ratios on the Strength and the Porosity

Figure 7.5 shows the compressive strength and the porosity of porous concrete in different acetone-to-toluene (A:T) ratios of 1:1 and 3:2. The coarse aggregate sizing 1/2": 3/8" was set to 50:50, applying from the test in the previous section. Similar results were found on the various w/c ratios as 0.30 provided the best conditions for both EPS admixture and the specific coarse aggregate used. Typical porous concrete strength was 12.02 MPa at their 28 days age, while the average porosity was 23%. Nevertheless, the maximum compressive strength and porosity were increased to reach the level of 17.89 MPa and 22.51%, respectively.

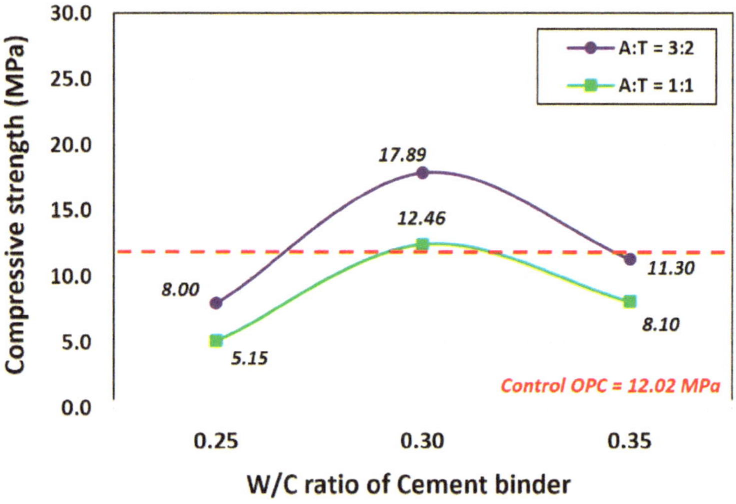

(a) Compressive strength of porous concrete

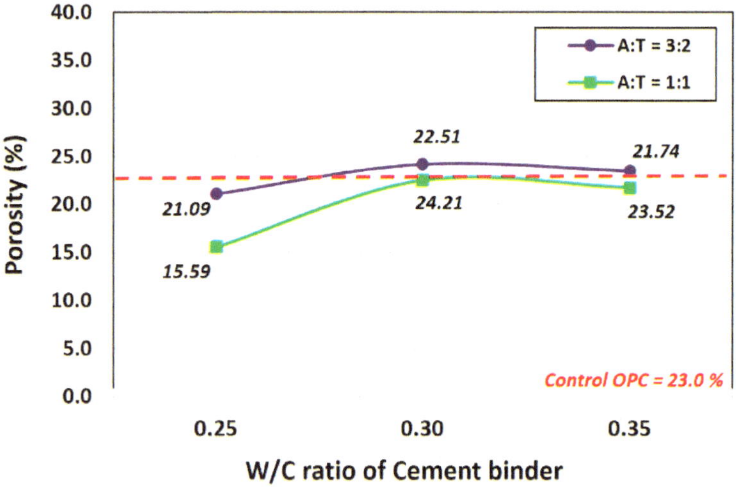

(b) Porosity of porous concrete

Fig. 7.5 Properties of porous concrete with different A:T ratios at 28 days of age

Fig. 7.6 Porous concrete specimens with EPS foam waste as an admixture

In general, acetone provides a hardening process for the mixture, but it has a brittle property. Toluene, typically, is a bit more cohesive and provides binding/bonding ability to the mixture. At the A:T of 1:1, a high portion of toluene provided quite an inert property as well as extended the setting time of concrete. While A:T of 3:2 achieved more workability with a suitable condition for the mixing process. For these reasons, the coating of the binder on the gravel's surface was thus improved and gained higher strength than the other one. The specimens and final products of EPS foam waste as an admixture are as shown in Fig. 7.6.

7.5 Conclusion

From the testing results, it can be concluded that the EPS foam wastes can be used as an alternative polymeric admixture to strengthen the cementitious binder of porous concrete. With 10% addition to the conventional cement paste, the compressive strength and porosity were better than that of typical porous concrete with no admixture. At these conditions, the appropriate water to cement (w/c) ratio was 0.30. The optimal coarse aggregate sizing 1/2":3/8" was 50:50 with the A:T ratio of 3:2. The maximum compressive strength and porosity were increased to reach the level of 17.89 MPa and 22.51%, respectively. The outcomes could not only improve the specific property for the porous concrete as construction materials but also promote the recycling of EPS foam wastes and allow society to be greener and cleaner.

Acknowledgements This research work was partially supported by Chiang Mai University. The authors would like to express gratitude to the Department of Civil Engineering, Faculty of Engineering, Chiang Mai University (CMU), for providing equipment and facilities. This research work was partially supported by Faculty of Engineering, Chiang Mai University, Thailand. Special thanks to my lovely and hardworking students, Chanwit Worapanyasakulchai, Natchanon Harnkoedsiri and Siwat Srithongin, who kindly helped to conduct the experimental works.

References

Ariffin NF, Jaafar MM, Lim NAS, Bhutta MAR, Hussin MW (2018) Mechanical properties of polymer-modified porous concrete. *In IOP Conference Series: Materials Science and Engineering* **342** 1 pp 012081. IOP Publishing

Bualuang T, Jitsangiam P, Suwan T, Rattanasak U, Jakrawatana N, Kalapat N, Nikraz H (2021) Non-OPC binder based on a hybrid material concept for sustainable road base construction towards a low-carbon society. J Mater Res Technol 14:374–391

Eskander SB, Saleh HM, Tawfik ME, Bayoumi TA (2021) Towards potential applications of cement-polymer composites based on recycled polystyrene foam wastes on construction fields: impact of exposure to water ecologies. Case Stud Const Mater 15:e00664

GISTDA (2019) The 5-years report of foam wastes. The Geo-Informatics and Space Technology Development Agency (Thailand). Accessed from http://coastalradar.gistda.or.th/wp/?page=news-detail&id=3561 on 01/01/2022

Jitsangiam P, Suwan T, Pimraksa K, Sukontasukkul P, Chindaprasirt P (2021) Challenge of adopting relatively low strength and self-cured geopolymer for road construction application: a review and primary laboratory study. Int J Pavement Eng 22(11):1454–1468

Maichin P, Suwan T, Jitsangiam P, Chindaprasirt P, Fan M (2020) Effect of self-treatment process on properties of natural fiber-reinforced geopolymer composites. Mater Manuf Process 35(10):1120–1128

Matar P, Barhoun J (2020) Effects of waterproofing admixture on the compressive strength and permeability of recycled aggregate concrete. J Build Eng 32:101521

Toghroli A, Shariati M, Sajedi F, Ibrahim Z, Koting S, Mohamad ET, Khorami M (2018) A review on pavement porous concrete using recycled waste materials. Smart Struct Syst 22(4):433–440

Wattanachai P, Suwan T (2017) Strength of geopolymer cement curing at ambient temperature by non-oven curing approaches: an overview IOP Conference Series. Mater Sci Eng 212(1):012014

Chapter 8
Building a Sustainable Municipal Solid Waste Treatment System in Japan—A Critical Review

Shuheng Zhao and Hiroshi Onoda

Abstract In this study, we looked at examples of local power producers and suppliers (local PPS) with waste-to-energy systems to understand the current role of the municipal solid waste (MSW) treatment system in building a sustainable society. We have also reviewed the process of centralizing MSW treatment facilities in line with the policy trends of the MSW treatment system in Japan. A case study and analysis of the Kakegawa regional circular and ecological sphere (regional CES) concept, which contributed to the construction of the Japanese version of Stadtwerke, focused on the MSW treatment system in Kakegawa City (Shizuoka Prefecture) was presented to demonstrate an example of a sustainable MSW treatment system. The case study was discussed via two perspectives: public–private partnerships (PPPs) and prioritizing local public interest. It was found that the involvement of the public in PPPs and the improvement of social acceptance by building the trust of the citizens by considering public interest are of utmost importance.

Keywords MSW treatment · Regional CES · PPPs · Local PPS · Stadtwerke · Waste to energy

8.1 Introduction

According to the materials from Japan's Ministry of the Environment, Japan's MSW treatment system has gone through the following three stages with the enactment of the legal system: (1) era for improving public health (1900 Waste Cleaning Law-until present); (2) era for pollution problems and conservation of the living environment (1970 Waste Management Act-until present); and (3) era for building a sound material-cycle society (1991 Waste Management Act revision-until present) (Ministry of the Environment 2014).

S. Zhao (✉) · H. Onoda
Graduate School of Environment and Energy Engineering, Waseda University, 513 Wasedatsurumakicho, Tokyo, Shinjuku-ku 162-0041, Japan
e-mail: zhaoshuheng@fuji.waseda.jp

J. Baeyens et al. (eds.), *Proceedings of 2022 4th International Conference on Environment Sciences and Renewable Energy*, Environmental Science and Engineering, https://doi.org/10.1007/978-981-19-9440-1_8

After the establishment of the "Sustainable Development Goals" (SDGs) at the United Nations Summit on September 25, 2015, Japan proposed the concept of "regional circular and ecological sphere" (regional CES) in 2018 called "local SDGs". Regional CES aims to maximize the vitality of the region by complementing and supporting resources according to the characteristics of the region while forming an independent and decentralized society to maximize the use of local resources such as the beautiful natural landscapes in each region (https://www.env.go.jp/seisaku/list/kyoseiken/index.html). This provides the future image of a sustainable society with the renewal of the Fifth Basic Environmental Plan for the promotion of a circular economy and the MSW treatment facility Development Plan, ushering in an era of a sustainable MSW treatment system.

In 1997, the Ministry of Health and Welfare implemented a wide-area waste management plan with the main purpose of reducing the emission of dioxins associated with waste disposal, etc. On March 29, 2019, the Ministry of the Environment released a notice on "Wide-area Waste Management and Concentration of MSW treatment facilities for Sustainable Treatment" in response to the new trends in waste disposal brought about by the decrease in the amount of municipal solid waste due to population decline and the promotion of the 3R Initiative, etc. In addition, as a measure to promote the expansion and consolidation of MSW treatment facilities, the concept of "private utilization" was presented in which municipalities can outsource waste treatment to private MSW treatment facilities (https://www.env.go.jp/hourei/11/000652.html).

The utilization of private funds and their expertise in the form of the so-called PPP is necessary to reduce the financial costs. The government can play a leading role with their knowledge on the operation of conventional MSW treatment systems, and giving back to the local public by creating new value in the region by positioning the MSW treatment facility as a regional energy center and social infrastructure in the city.

In this study, we looked at examples of regional energy centers focused on the MSW treatment system to understand the current state of the MSW treatment system in building a sustainable society while organizing the policy trends of Japan's MSW treatment system. We also presented a case study and analysis of the Kakegawa regional CES concept, which contributed to the construction of a sustainable MSW treatment system in Kakegawa City (Shizuoka Prefecture) from the two perspectives of stakeholder management in PPP and giving priority to the local public interest. Finally, we presented the direction toward the realization of a sustainable MSW treatment system in Japan.

8.2 Methodologies

In this study, to understand the current status and issues of the MSW treatment system in building a sustainable society, we collected based on previous literature existing cases that contribute to the construction of a regional energy center focused on the

MSW treatment system. Simultaneously, we clarified the policy trends in the MSW treatment system that accompanies the legal system. For a case study, we examined the Kakegawa regional CES concept, which contributed to the construction of a sustainable MSW treatment system in Kakegawa City (Shizuoka Prefecture).

The data required for the analysis was collected through various websites and reports, minutes of the city council, articles of the council, literature surveys such as newspaper articles, and field surveys such as hearings and interview surveys.

8.2.1 Constituents of MSW

In Japan, "industrial waste" refers to 20 types of waste generated by business activities as defined by law, while other wastes fall under "MSW". MSW mainly consists of (1) household waste (combustible waste, noncombustible waste, large-size waste, four types of home appliances (air conditioners, televisions, refrigerators, washing machines), personal computers, automobiles, and hazardous waste), and (2) business-related MSW (combustible waste, large-size waste from businesses other than industrial waste), and (3) human waste (human waste and sludge from septic tanks) (https://www.kankyo.metro.tokyo.lg.jp/resource/general_waste/about.html).

The municipalities handle the disposal of the above MSW. In addition, garbage is sorted by the waste generators. The rules for sorting vary from municipality to municipality, but the main rules are as follows: (1) recyclable resources (recyclable paper, plastic containers and packaging materials, bottles, cans, PET bottles, spray cans, cassette gas cylinders, batteries), and (2) combustible waste (plastic items except for plastic containers and packaging materials, kitchen waste, cooking oil, clothes, disposable diapers, paper scraps, small amounts of branches from garden plants, rubber/leather items), and (3) metal, ceramic, glass (metal, ceramic, glass, small home appliances) (https://www.city.shinjuku.lg.jp/seikatsu/file09_02_00001.html). According to the Home Appliance Recycling Law, household air conditioners, televisions, refrigerators, washing machines, classified as four types of home appliances, must be sent for recycling by contacting the Home Electric Appliance Reception Center. In addition, furniture, bicycles, and other household waste exceeding 30 cm in length are classified as large-size waste. Prior to disposal, an appointment must be made with the Large-Sized Waste Reception Center (https://www.city.shinjuku.lg.jp/seikatsu/file09_02_00001.html).

8.3 Current Status of the MSW Treatment System

8.3.1 Consolidation of MSW Treatment Facilities

According to the data from Japan's Ministry of the Environment, the total amount of waste discharged nationwide in 2019 was 42.74 million tons, and the total amount of waste treated was 4,095 tons, excluding the total collection amount (https://www.env.go.jp/recycle/waste_tech/ippan/stats.html). Figure 8.1 shows the changes in the total volume of waste disposal nationwide by year. The total amount of waste disposed of gradually decreased for nine consecutive years from 2010 to 2018. However, the total amount of household waste increased slightly in 2019 due to the influence of telework and other measures to prevent the spread of the COVID-19 pandemic. Meanwhile, the amount of waste that is directly incinerated accounted for more than 80% of the total amount of waste that had been continuously treated over the years. Figure 8.2 shows the changes in the number of MSW treatment facilities by scale. The number of MSW treatment facilities gradually decreased from 2010 to 2019 for ten consecutive years. Despite the decrease in the number of small MSW treatment facilities with a processing capacity of 50 tons or less per day, the number of MSW treatment facilities with a processing capacity of 100 tons or more per day is gradually increasing, indicating a tendency to consolidate MSW treatment facilities.

On the other hand, looking at the results of the survey on the consolidation of MSW treatment facilities for each regional government conducted between November and December 2019, measures such as "private utilization" are well recognized. However, it is more common to focus on reducing fiscal costs, and awareness of the new value creation in the region is still insufficient (Ogawa and Onoda 2020).

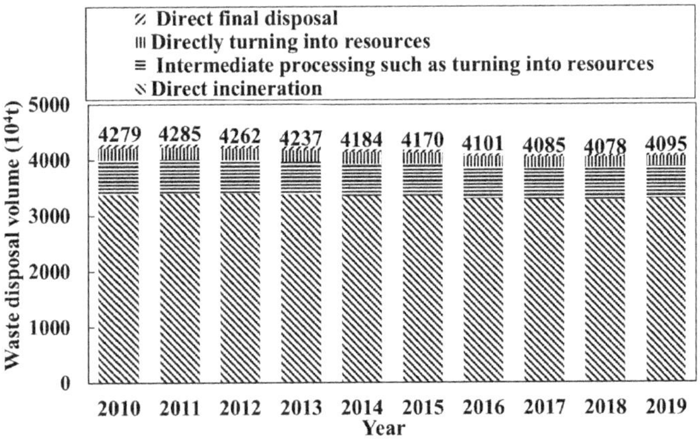

Fig. 8.1 Changes in waste disposal volume in Japan. Data obtained from https://www.env.go.jp/recycle/waste_tech/ippan/stats.html

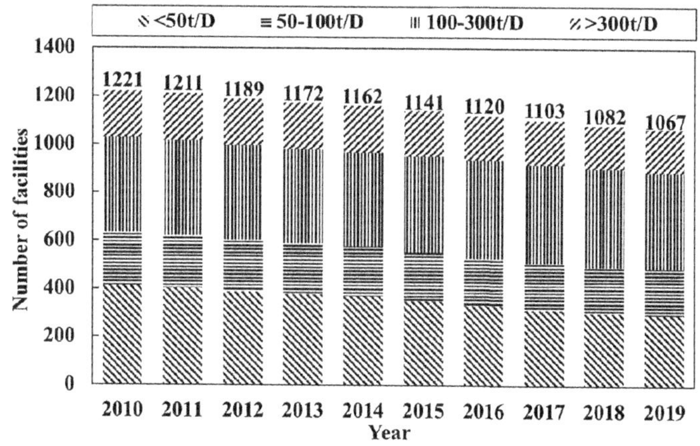

Fig. 8.2 Changes in the number of MSW treatment facilities in Japan. Data obtained from https://www.env.go.jp/recycle/waste_tech/ippan/stats.html

8.3.2 Utilization of Public–Private Partnership/Private Finance Initiative (PPP/PFI)

According to the MSW treatment business fact-finding survey, the ratio of utilizing PPP/PFI in MSW treatment facilities exceeded 60% in 2016 (https://www.env.go.jp/recycle/waste_tech/ippan/stats.html). Most of them were to reduce the financial burden due to soaring costs in the construction and maintenance of new MSW treatment facilities due to the deterioration of MSW treatment facilities. Especially in relatively small municipalities, while human resources and financial constraints are increasing, the decrease in waste emissions due to the declining population is remarkable. Recently, the conventional MSW treatment facilities directly managed by the regional government have been shifting to maintenance and operation mainly by private companies such as businesses accepting industrial waste. Moreover, there are cases of processing of municipal solid waste and industrial waste in facilities of so-called PFIs, and two specific examples are given here.

(1) A PPP agreement to promote regional energy center development and operation projects in Aioi City (Hyogo Prefecture)

On October 4, 2021, Aioi City (Hyogo Prefecture) (population: 28,000, as of 2021 (https://www.city.aioi.lg.jp/soshiki/kankyo/sinbika.html)) began to reduce the financial burden by promoting the regional energy center maintenance and operation project as a new MSW treatment facility. They signed a PPP agreement with private companies such as industrial waste companies, environmental consulting companies, and environmental plant manufacturers, to resolve regional waste issues, including dealing with disaster waste and marine debris (https://www.city.aioi.lg.jp/soshiki/kankyo/sinbika.html). The objective is to build a regional CES centered on renewable energy from waste by establishing a Special Purpose Company (SPC) for facility maintenance and operation via a PPP. This setup can utilize private funds and expertise while still maintaining public involvement. In addition, through this setup, the electric power and thermal energy generated from municipal solid waste inside and outside the city, as well as industrial waste such as construction waste approved by Aioi City (stopping delivery in the event of a disaster), are utilized effectively.

(2) A memorandum of understanding for the promotion of MSW treatment facility maintenance and operation business in five towns in Kamimashiki (Kumamoto Prefecture)

On October 1, 2021, five towns in Kamimashiki District in Kumamoto Prefecture, including Mifune, Kashima, Mashiki, Kosa, and Yamato (population: 82,000, as of 2021 (https://www.town.mashiki.lg.jp/)), signed a memorandum of understanding for the promotion of the maintenance and operation business of an MSW treatment facility to be constructed in Mifune Town to proceed the discussions with private companies, industrial waste companies, and local waste companies. This resulted in a reduced financial burden on the regional government by lowering the initial cost from 13.9 billion yen to 1.2 billion yen by renting out the land to private entities (https://www.dinsgr.co.jp/news_no200/). In addition, it is expected that disaster waste and industrial waste, as well as municipal solid waste (within the five towns in Kamimashiki), will be treated and incinerated. They also aim to build a regional CES centered on MSW treatment facilities by converting waste generated in the region into energy and returning it to the region.

In areas where MSW emissions are declining, as described above, there is a large surplus of incineration capacity (about 44,700 t/D (Takaoka 2020) nationwide) due to the decrease in MSW disposal and inefficiencies in waste disposal are becoming apparent. In contrast, co-firing MSW with industrial and disaster waste promotes efficiency in waste treatment and secures waste generation. For example, the Mie Recycling Center, an industrial waste treatment facility, processes approximately 160,000 tons of MSW and 440,000 tons of industrial waste per year (it also accepts disaster waste). In addition, it has a power generation capacity of 1.5 kW (Yoshidome et al. 2021). Because of this, there are expectations that a stable supply of waste will be secured by accepting industrial waste at the MSW treatment facility. It is also expected that it will serve as a regional infrastructure facility that improves regional resilience by stopping the delivery of industrial waste in the event of a disaster and preferentially treating regional disaster waste (Minamikawa 2016). In addition, the government's involvement in monitoring the operational status of the regional facility during the project period, such as deciding which industrial waste are to be accepted in the agreement, is required to have a PPP approach that promotes the building of regional trust relationships. This is done by gaining the understanding of residents in the location of treatment facilities (Hata 2022). In addition, when cooperating with an industrial waste treatment facility, it is necessary to design an appropriate scale in anticipation of a decrease in the amount of waste (Osako 2022). The above two cases have taken a step toward the search, but since both are planned projects, the key point is how to manage the realization of such a PPP system in the future.

8.3.3 Waste Power Generation and Local Power Producers and Suppliers (Local PPS)

According to the MSW treatment business fact-finding survey, the amount of power generated per waste treatment amount gradually increased from 206 kWh/t in 2010 to 292 kWh/t in 2019 due to the consolidation of MSW treatment facilities (https://www.env.go.jp/recycle/waste_tech/ippan/stats.html). In addition, as of 2019, the total number of MSW treatment facilities is 1,067. 384 facilities (about 36%, of which valid responses are 381) had power generation facilities, and the total power generation capacity was 2,078 MW (https://www.env.go.jp/recycle/waste_tech/ippan/stats.html). Figure 8.3 shows the number of MSW treatment facilities with power generation facilities by year for each power generation capacity. The total number of MSW treatment facilities with power generation facilities is gradually increasing, but most of them tend to be concentrated in 2,000–5,000 kW MSW treatment facilities.

While there are many relatively small facilities of less than 2,000 kW, the joint installation of MSW treatment facilities by multiple regional governments is also progressing. Although the amount of waste generated is decreasing as the facilities tend to become larger, the number of power generation facilities and the total power generation capacity are increasing. Therefore, waste power generation has the potential as a power source for regional energy centers (Inagaki 2017). In addition, waste power generation utilizes a low-carbon power plant and has the potential as a relatively stable base load power source that is not affected by weather conditions. This highlights the role of waste power generation as one of the distributed power sources in the regional energy business (Japan Environmental Sanitation Center 2017).

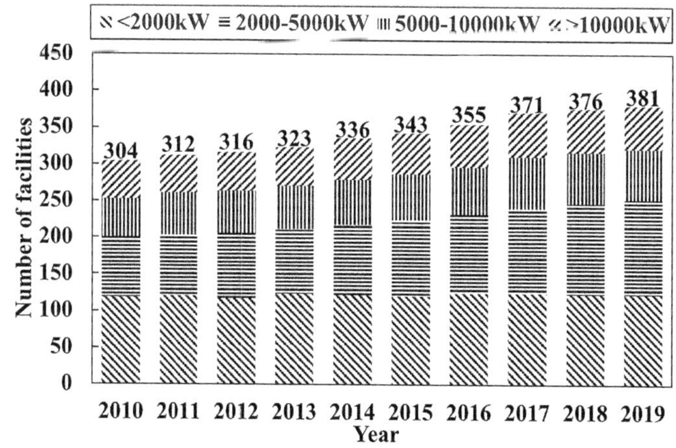

Fig. 8.3 Changes in the number of MSW treatment facilities with power generation facilities in Japan. Data obtained from https://www.env.go.jp/recycle/waste_tech/ippan/stats.html

On the other hand, as the use of MSW treatment facilities expands, it is important to keep in mind the pollution problems caused by waste incineration. In Japan, dioxin emissions at MSW treatment facilities have been contained from 5000 g-TEQ/Y in 1997 to 20 g-TEQ/Y in 2019 (Ministry of the Environment 2021a). According to a survey conducted from April 2019 to April 2020 measuring dioxin concentrations in MSW incineration facility flue gas, only four of the 1843 MSW incinerators in Japan exceeded emission standards. This finding resulted in two being decommissioned and two being repaired to comply with the standard levels before the operation is resumed (Ministry of the Environment 2021b). While carefully implementing measures, there is a need to promote the expansion of the use of MSW treatment facilities.

With the start of full liberalization of electricity retailing in April 2016, a total of 727 power producers and suppliers (PPS) have been established throughout Japan as of the end of June 2021 (https://www.enecho.meti.go.jp/category/electricity_and_gas/electric/electricity_liberalization/what/). However, while the number of PPSs continues to grow, there were 95 cases of business succession and 38 cases of business abolition/dissolution recorded. This was due to the problem of soaring market electricity prices during the winter season (December to January) (https://www.meti.go.jp/shingikai/enecho/denryoku_gas/denryoku_gas/pdf/037_03_00.pdf). There has been a growing concern that such focus on electric power can cause PPSs to become self-serving, engendering little price competition, and contributing to new regional monopolies (Zhao and Onoda 2021). Among the PPSs, those that have power generation limited to a single area and mainly supply power to public facilities, private companies, and homes in that particular area are called "local PPS". As of 2017, there are about 162 companies categorized as such (Trade and Industry 2017). Local PPS can be divided into (1) companies with regional government as an investor; (2) companies with indirect involvement from the regional government; and (3) companies led solely by a private company. The first type is also called the "regional government PPSs" company, and it is expected that the government will play a leading role in the operation of the business entity. As of 2020, there are 44 regional government PPSs (Zhao and Onoda 2021). Out of this 44, 39 companies have disclosed their power source composition, 8 of which are using waste power generation accounting for about 20.5% of those that disclosed power source composition (Table 8.1).

From Table 8.1, it can be seen that waste power generation is used as the main power source, but more than half of it is used in combination with other power sources (solar power, hydropower, wind power, biomass). Also, looking at the population size of the regional government, they range from small ones of 100,000 or less (2 cases) to large local PPSs of 200,000 to 1 million (4 cases) and 1 million or more (2 cases). Furthermore, while most local PPSs are sold to public facilities as a whole, some also include ordinary households.

Table 8.1 List of new waste-to-energy local power producers and suppliers funded by regional governments

Prefecture	Company name	Renewable energy sources	Sales destination	Capital stock (yen)/regional governments investment ratio (%)
Saitama	Chichibu Power Supply, Inc.	Waste to energy, megasolar, hydropower	Public facility, private facility	20 million/95
Chiba	Narita Katori Energy, Inc.	Waste to energy, megasolar	Public facility	9.5 million/80
Niigata	Niigata Swan Energy, Inc.	Waste to energy, megasolar, biomass	Public facility	50 million/10
Nagano	Marubeni Ina Power, Inc.	Waste to energy	Public facility	50 million/10
Hiroshima	Fukuyama Mirai Energy, Inc.	Waste to energy, megasolar, biomass, hydropower	Public facility	100 million/10
Shizuoka	Hamamatsu Power Supply, Inc.	Waste to energy, megasolar	Public facility, private facility	60 million/8.3
Fukuoka	Kitakyushu Power, Inc.	Waste to energy	Public facility, private enterprise	60 million/24.2
Kumamoto	Smart Energy Kumamoto, Inc.	Waste to energy	Public facility	100 million/ 5

Source Prepared by the author based on https://chichibu-pps.co.jp/, https://www.nk-energy.jp/, https://niigata-se.co.jp/, https://www.ina-mirai-denki.com/, https://fukuyama-miraienergy.co.jp/, https://www.hamamatsu-e.co.jp/, https://se-kumamoto.co.jp/

Here, we will introduce the cases of Kitakyushu Power Inc. (Fukuoka Prefecture) and Smart Energy Kumamoto Inc. (Kumamoto Prefecture), which use waste power generation as the main power source.

(1) Kitakyushu Power Inc.

In 2013, Kitakyushu City (population: 930,000, as of 2021) aimed to supply low-carbon, stable, and inexpensive energy, and the "Kitakyushu City Regional Energy Base Promotion Project" was implemented as a major project for the new growth strategy (https://www.city.kitakyushu.lg.jp/). As one of these efforts, in December 2015, Kitakyushu City invested 24.2% and established a local PPS company, Kitakyushu Power Inc., jointly with an electric appliance manufacturer, an IT company, and a private company of a local financial institution. In April 2016, it started selling electricity to local consumers such as public facilities and private companies, mainly in the city and reached about 200 facilities as of 2021 (https://kitaqpw.com/). Kitakyushu City is implementing a wide-area acceptance of municipal solid waste from other cities; the mechanism is shown in Fig. 8.4. In the Kitakyushu area, Kitakyushu City, the core city of the region, accepts garbage from the surrounding cities and towns of Yuihashi City, Miyako Town, Onga District, Nakama City, and Nogata City, and the processing is carried out at three MSW treatment facilities in the city. The waste treatment capacity and power generation capacity of each facility are 720 t/D/23,500 kW, 600 t/D/6,000 kW, and 810 t/D/17,200 kW, for a total of 46,700 kW. In addition, when receiving garbage, Kitakyushu City has the role of adjusting the carry-in from the city and surrounding cities and towns according to the disposal status of the MSW treatment facilities at three locations in the city. In Kitakyushu City, the facilities to which energy is supplied will be gradually expanded according to the amount of waste generated at the city's MSW treatment facility, and by 2025, they are aiming for 100% electrification of all public facilities (about 2,000 facilities) in the city by the electricity coming from the city's renewable energy power (waste power, solar power, wind power, biomass) plant. Furthermore, they are promoting a power storage system that solves various issues in the expansion of renewable energy by effectively utilizing storage batteries as a development strategy for the "100% Renewable Energy Kitakyushu Model" in the future (https://kitaqpw.com/). By storing the power generated at the city's renewable energy power when the cost of power is low and discharging it to the facility when the cost is high, they are aiming to create value in the region by effectively utilizing surplus electricity and reducing electricity prices, thereby achieving the act of giving back to the local public promoting the public interest.

(2) Smart Energy Kumamoto Inc.

As part of the promotion of the "Kumamoto City Low Carbon City Strategic Development Plan" enacted in 2015, the "Kumamoto City Earthquake Reconstruction Plan" enacted in 2016, and one of the initiatives for local production for local consumption of waste energy, Kumamoto City (population: 738,000, as of 2021 (https://www.pref.kumamoto.jp/)) and JFE Engineering, a private enterprise plant maker, established a local PPS company, Smart Energy Kumamoto Inc., in November 2018. Kumamoto City and JFE Engineering have a 5% stake and 95% stake, respectively. In May 2019, the company started supplying energy to major public facilities such as ward offices, government buildings, and schools, and as of 2021, it has supplied about 40% of the city's public facilities in 220 locations (https://se-kumamoto.co.jp/). In Kumamoto City, power is supplied by renewable energy (waste power) mainly from the power generated by the two MSW treatment facilities in the city (Fig. 8.5). The waste treatment and power generation capacity of the western and eastern MSW treatment facilities are 600 t/D/10,500 kW and 280 t/D/5,980 kW, respectively, for a total of 16,480 kW. In addition, the heat generated from the MSW treatment facility is supplied to nearby farms, disaster prevention centers, and hot spring facilities. Furthermore, as a disaster prevention measure, a large-storage battery of 704 kWh

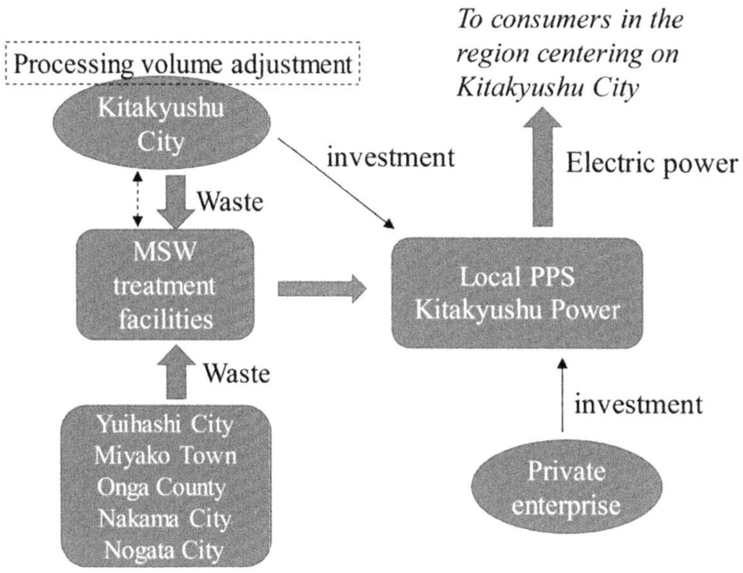

Fig. 8.4 Image of waste distribution adjustment and electricity use in the Kitakyushu region. *Source* Prepared by the author based on https://kitaqpw.com/

was installed in the water and sewage government building, and 588 kWh was installed in the ward office. The battery plays an important role in the event of a disaster. Moreover, by implementing measures to secure electricity in the event of a disaster, such as laying a private line, a regional value is created, and giving back to the local public is attained.

Using the two examples, it was demonstrated that in relatively large cities, the public plays a leading role in the management of waste-to-energy businesses, and the local PPS companies that use waste power as their main power source are doing relatively well. As one of the risk diversification measures, the government is taking the lead, creating regional values through PPPs, and building relationships of trust with citizens by giving back to the local public, in addition to providing electricity to the community.

8.4 Trends in MSW Treatment Systems Overseas

In this section, we have collected examples of MSW treatment systems in Germany, the United Kingdom, and China to understand the trends in MSW treatment systems overseas, especially on the business methods (Table 8.2).

Overseas, MSW treatment facilities that are relatively larger than those in Japan are often seen, and most of them have power generation facilities. From the perspective of a business entity, a unique business method different from Japan is adopted, such

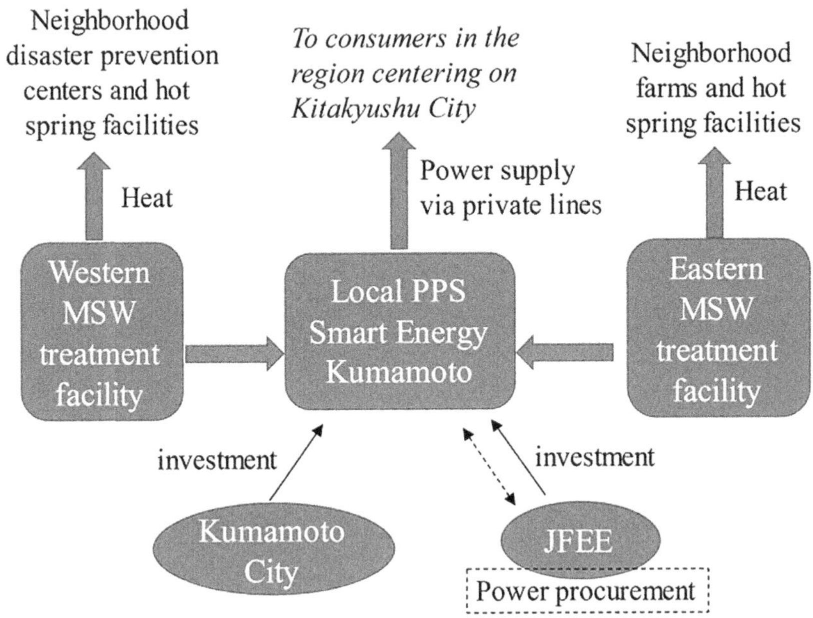

Fig. 8.5 Image of waste distribution adjustment and electricity use in the Kumamoto region. *Source* Prepared by the author based on (https://se-kumamoto.co.jp/)

as a public corporation type by a city public corporation, LA/Merchant by a private company, and PFI build-operate-transfer (BOT) by a state-owned enterprise.

In Germany, the incinerator rate was 35% as of 2014 (https://ec.europa.eu/eurostat/documents/2995521/7214320/822032016-APEN.pdf/eea3c8df-ce89-41e0-a958-5cc7290825c3). The same situation was predicted in Japan, where the capacity of the MSW treatment facility seems to be excessive with respect to the amount of waste. As a result, the operation of the facility will be difficult, so the trend in Germany can be used as a reference (Chichibu 2015).

The Ruhleben MSW treatment facility in Berlin is owned by BSR (Berliner Stadtreinigung), a city cleaning corporation wholly owned by the city of Berlin. It processes waste from 3.5 million people in the city, for a total of 60% of the total annual waste in Berlin, which is 533,000 tons of municipal solid waste. Power generation facilities are not maintained themselves. The generated steam will be sent to the adjacent coal-fired power plant, which will have an annual power generation capacity of 180 GWh, and will be able to supply electricity and heat to 5% of the households in Berlin (https://www.bsr.de/). In addition, BSR provides not only MSW treatment services but also a wide range of public services such as vehicle rental, thus creating value for the local public. An entity such as a social enterprise pursuing the maximization of local public interest is called Stadtwerke. Stadtwerke is a community-based business entity responsible for the development and operation of a wide range of regional energy and living infrastructure such as electricity,

Table 8.2 List of MSW treatment facilities overseas

Region (country)	Regional population (people)	Scale (t/Y)	Power generation capacity (GWh/Y)	Business method	Division of roles between public and private sectors
Berlin (Germany)	3,500,000	533,000	180	Public	Facility construction and operation, maintenance, financing (cleaning corporation)
Oxfordshire (UK)	688,000	326,000	218	LA/ Merchant	Conclusion of waste supply plan (administrative), Facility ownership, operation management, financing (private company)
Guangzhou (China)	18,740,000	2,800,000	1,300	PFI (BOT)	Supervision (administration), facility construction and operation, maintenance, financing (state-owned enterprise)
Wuzhou (China)	2,821,000	470,000	164	PFI (BOT)	Supervision (administration), facility construction and operation, maintenance (SPC), financing (private)

Source Prepared by the author based on https://www.berlin.de/, https://oxfordcity.co.uk/, http://www.gz.gov.cn/, http://www.wuzhou.gov.cn/, https://www.bsr.de/, https://www.viridor.co.uk/energy/energy-recovery-facilities/ardley-erf/, http://www.gxcic.net/subject/ztbd/showgjgc.asp?ID=173519. LA = local authority; BOT = build-operate-transfer; SPC = Special Purpose Company

gas, heat, water, transportation, communication, and public facility management in various parts of Germany. In recent years, re-nationalization has progressed, and there are about 1,000 companies throughout Germany (Wuppertal Institut 2018), the essence of which is "maximizing the local public interest" (Onoda 2018). Japan's regional government does not have the expertise and customer base to operate an energy business, so it is quite a hurdle to apply Germany's experience to Japan as it is. However, by utilizing the expertise on the operation of an MSW treatment system owned by the conventional government, and working on not only the energy business of waste power generation but also the development and operation of a wide range of regional infrastructure and the provision of public services through PPP, it is possible to develop a Japanese version of Stadtwerke centered on the MSW treatment facility that builds regional trust by creating regional values.

In the United Kingdom, PFI contracts were terminated in 2013 due to a decrease in business income as the amount of waste decreased and as a result of the opposition from environmental groups and residents (https://www.env.go.jp/recycle/report/r3-08/co2.html). A private company such as a major waste business company (Merchant) will finalize a waste supply contract with a local authority (LA) to utilize private funds and expertise to construct an MSW treatment facility giving rise to the company called LA/Merchant. Ardley Energy Recovery Facility (ERF) in Oxfordshire, which acquired LA/Merchant, is owned by the major waste business company Viridor. They carry in and accept municipal solid waste from Oxfordshire County, as well as industrial waste discharged from regional governments and companies that have finished other contracts. They also process 326,000 tons of waste annually and have an annual power generation capacity of 218 GWh, which creates regional value by providing electricity to about 60,000 households, thereby building a relationship of regional trust (https://www.viridor.co.uk/energy/energy-recovery-facilities/ardley-erf/).

In China, the use of PFI (BOT) in the field of MSW treatment is widespread, but as in Japan, differences due to regional scale are common. For example, the large-scale Fushan Circular Economy Trade and Industry Park (MSW treatment facility) in Guangzhou has the largest scale in China (2.8 million tons/Y, 1,300 GWh/Y) (https://www.askci.com/xmal/20190624/1203301148258.shtml), and there is a perfect balance between the city scale and private companies. In this case, the Guangzhou Environmental Protection Investment Group Company, which is a state-owned enterprise, serves as a private company. On the other hand, there is also an example from the Wuzhou Circular Economy Industrial Park (MSW treatment facility) in Guangxi, which is relatively small, but carries in and accepts municipal solid waste in the city, as well as industrial waste such as medical waste, disposes of 470,000 tons/Y of waste, and maintains a power generation capacity of 164 GWh/Y. In this case, an SPC, which is the main maintenance and operation entity, was established in the form of a joint investment with a private company (10% city, 90% private sector) (http://www.gxcic.net/subject/ztbd/showgjgc.asp?ID=173519). Furthermore, by establishing a jointly-funded SPC, they are trying to stabilize the business while balancing the appropriate risks and benefits of the public and private sectors. Nevertheless, in China, all initiatives on risk-sharing in PPPs are guidelines for each sector, and there is still no law that integrates them (Liu 2021).

Most of the "large-scale" MSW treatment facilities that cater to such "large" customers are located in the form of "industrial parks," which are like an aggregation of resource recycling system functions, with incineration facilities located on the same site as chemical plants and other facilities. Therefore, electricity and heat generated at the incineration facility can be directly utilized at the facility on the same site, enabling mutual coordination among the facilities in the industrial park. On the other hand, in Japan, where the scale of facilities is relatively smaller than in China, there is no need to pursue "large scale" merely by force, but with the premise of "the right person in the right place," the creation of such business schemes is required (Onoda 2022).

8.5 Kakegawa Regional Circular and Ecological Sphere Concept

The regional CES concept presents the direction in which the sustainable MSW treatment system, the so-called Japanese version of Stadtwerke, should aim for based on the above trends in the MSW treatment system in Japan and overseas. This chapter describes the Kakegawa regional CES concept claimed by Kakegawa City (Shizuoka Prefecture) as a case study.

8.5.1 Overview of Kakegawa City

Kakegawa City is a city located in the western part of Shizuoka Prefecture. In April 2005, (former) Kakegawa City, Daito Town, and Osuga Town merged (Fig. 8.6). As of 2021, the city area is 265.69 km^2 and the population is 116,907 (Table 8.3). In 2010 and 2011, it became the number one city in Japan for waste reduction for the second consecutive year. Until 2017, it continued to be the second city for waste reduction in Japan and reduced the amount of waste discharged per person per day to 637.2 g, compared to the national average of 920 g (https://www.city.kakegawa.shizuoka.jp/).

In Kakegawa City, there was a waste disposal problem due to the deterioration of the MSW treatment facility. On June 5, 2021, a "Declaration of Waste Management Emergency" was issued due to the breakdown of the incineration facility at the Environmental Resource Gallery (140t/D), an MSW treatment facility in Kakegawa City, and the neighboring Kikugawa City (population: 48,000, as of 2021 (https://www.city.kikugawa.shizuoka.jp/)) in Kakegawa City (https://www.city.kakegawa.shizuoka.jp/). Furthermore, on August 16 of the same year, a fire broke out in the Environmental Resources Gallery. This was presumed to be caused by an explosion of a lithium-ion battery, which burned down the facility and suspended the disposal of oversized garbage (combustible/non-combustible) and non-burnable garbage (http://www.kankyoshigen-gallery.jp/aboutus/index.html). In response to the current situation where the lack of sustainability is becoming apparent in the MSW treatment system, the city recognizes that it is extremely difficult to extend the life of the existing facility by improving the core. Thus, the city is now considering the development of a new MSW treatment facility. Incineration of garbage has the advantage of hygienic garbage disposal, especially kitchen waste. However, as mentioned earlier,

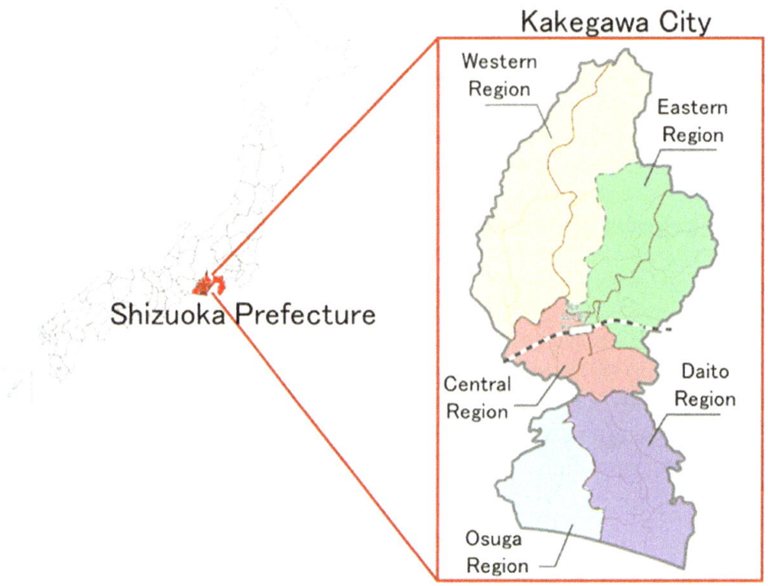

Fig. 8.6 Geographical location of Kakegawa City

Table 8.3 Socioeconomic overview of Kakegawa City (2020)

Population/number of households	116,907/45,794
Area (km^2)	265.69
Total production[a]	661.7 billion yen (6.04 billion USD)
Industrial composition[a] (primary industries: secondary industries: tertiary industries)	7: 40.3: 52.7

[a]Data from 2015 (https://www.city.kakegawa.shizuoka.jp/)

there is a risk of fire due to contamination of lithium-ion batteries, so it is necessary to establish a system to collect batteries in advance and to expand awareness of the need for sorting.

Furthermore, with the declining birthrate and aging population, the problem of garbage disposal by elderly people requiring long-term care is emerging not only in Kakegawa City but also in Japan on a nationwide scale. Since it is difficult for people requiring long-term care to transport garbage to the garbage collection point, the introduction of a garbage collection system such as automatic garbage collection is being considered. Meanwhile, in response to the spread of the COVID-19 pandemic on a global scale, a post-COVID-19 approach is required in the field of waste management for the medium to long term. Because of this, there is often a regional need to introduce a non-contact waste collection system (Onoda 2020).

8.5.2 Local PPS: Kakegawa Hotoku Power Inc.

In 2019, Kakegawa City conducted a feasibility study survey of a local PPS project by utilizing the Ministry of Economy, Trade and Industry's "Energy Structure Advancement/Conversion Understanding Promotion Project Cost Subsidy". On July 7, 2021, a local PPS company, Kakegawa Hotoku Power Inc., was established to solve regional issues brought about by the development of community businesses. This will be done through local production of electricity from solar and wind power for local consumption, regional economic circulation that creates a flow of funds by keeping part of the expenses that have flowed out of the city to the city, and reinvestment of profits from energy businesses in the development and operation of regional infrastructure (Kakegawa City 2021a). The total capital is 29.9 million yen, and the investors in descending order of investment rate are as follows: Kakegawa City, 34%; electric power companies (Pacific Power Co. Ltd., 10%; Japan Wind Development Co. Ltd., 10%; Kakegawa Furusato Energy Creation Co. Ltd., 1.7%); water and sewage company (Water Agency Inc., 10%); local waste company (Chuen Environmental Conservation Co. Ltd., 10%); industrial waste business-affiliated think tank (Daiei Kankyo Souken, 8%); financial company (NEC Capital Solutions Ltd., 7%); environmental nonprofit organizations (Earthlife Network, 3%; PV Network, 2%; Ohisama to Machizukuri, 0.3%); construction consultant (Showa Sekkei Co. Ltd., 3%); and gas companies (Chuen Gas Co. Ltd. of the Shizuoka Gas Group, 1%) (https://kakegawa.de-power.co.jp/). In addition, from April 2022, the power generated by five solar power plants installed in public facilities in the city and six wind power plants installed in the coastal areas will be used to supply power to all public facilities (70 locations) in the city (Kakegawa City 2021b). In the future, they will introduce waste power generation by using a new MSW treatment facility and are preparing to expand the supply to city businesses and general households, and to purchase the power generated by the household solar power generation system.

At Kakegawa Hotoku Power Inc., the private sector and the government will work together while involving stakeholders who are providers of public services, which include not only electric power companies, but also private companies with expertise on the development and operation of a wide range of regional infrastructure. This is to improve the productivity of the entire organization with synergistic effects such as unification between fields, information and resources sharing, and a role in providing a regional platform. The profits obtained from the energy business are not distributed to the investors, but instead are focused on the need to realize a "smart life" in which functions and services in all fields such as work, transportation, and medical welfare are linked by developing a community business to solve regional issues through PPPs. The business also has the following considerations: infrastructure optimization aimed at maintaining and updating infrastructure that can be used for future generations, energy saving through mutual energy interchange, energy optimization that allows local energy to be produced and consumed locally, and the environment (Fig. 8.7) (Kakegawa City 2021c).

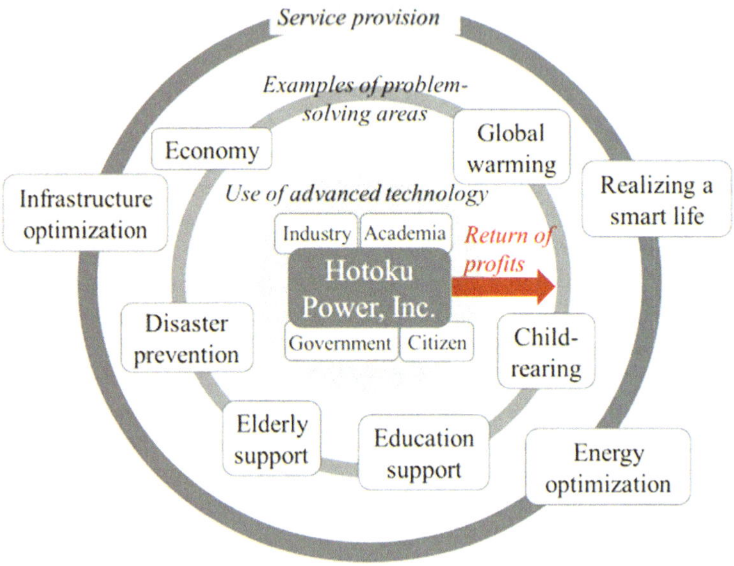

Fig. 8.7 Overview of the local power producer and supplier in Kakegawa city. Prepared by the author based on Kakegawa City (2021c)

8.5.3 Kakegawa Regional Circular and Ecological Sphere Concept Centered on MSW Treatment Facilities

The existing MSW treatment facilities in Kakegawa city are aging. In addition, it is expected that the amount of waste discharged will decrease due to the declining birthrate and aging population, while a large amount of disaster waste is expected to be generated after the Nankai Trough earthquake. Therefore, they are considering the construction of a new MSW treatment facility. To reduce the large financial burden for the total cost of developing and operating a new MSW treatment facility, which is estimated at 46 billion yen, Kakegawa City will set up a partnership agreement between the public and the private sector. After which, it will adopt a PPP method of PFI, in which the public entrusts waste disposal to facilities maintained and operated by the private sector with their funds, and conduct a feasibility study of the project. If it turns out that there is business potential, a business agreement will be finalized between the public and the private sector (http://www.kankyoshigen-gallery.jp/aboutus/index.html).

Figure 8.8 shows the maintenance and operation mechanism (basic concept) of a new MSW treatment facility. By establishing an SPC while utilizing private funds and expertise, they aim to solve local waste problems such as the generation of marine debris and disaster waste. At the same time, it is possible to flexibly respond to changes in the amount and quality of waste due to population decline, etc. by carrying in and accepting industrial waste in addition to municipal solid waste (stopping when

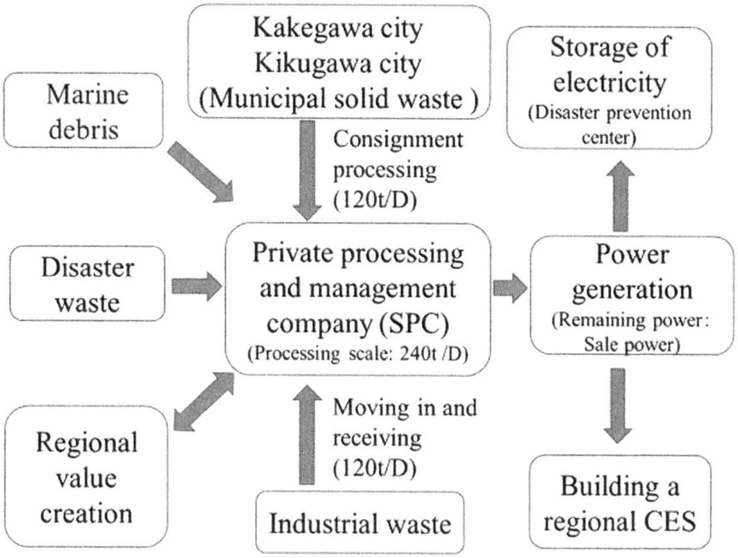

Fig. 8.8 Diagram showing the facility maintenance of the new MSW treatment facility. Prepared by the author based on http://www.kankyoshigen-gallery.jp/info/pdf/20210330document01.pdf

a disaster occurs). By doing so, they will create a sustainable MSW treatment system while generating regional value (http://www.kankyoshigen-gallery.jp/info/pdf/20210330document01.pdf).

In addition, Kakegawa City plans to build a Kakegawa regional CES in collaboration with the local PPS, and with the new MSW treatment facility as a community business base (Fig. 8.9). The electricity generated at the MSW treatment facility will be supplied via the local PPS system and private line, and the surplus electricity will be used to promote the use of electric vehicles (EVs). Meanwhile, the generated heat is supplied to the facility and stored as a disaster prevention measure. In addition, Kakegawa City established a fund to utilize the environmental purpose tax by bringing in industrial waste from outside the region. As one of the demonstrations and business financial resources, they aim to give back to the local public by supporting the development of community businesses. From here, we will introduce a waste disposal support service as a concrete example of community business development in the waste field.

Kakegawa City needs a new garbage collection service to deal with the increase in self-neglect (decrease in motivation for sorting and taking out the garbage) due to the declining birthrate and aging population, and social changes caused by COVID-19 (noncontact garbage collection). In response to the difficulty in increasing tax revenue and securing stable financial resources, reinvesting revenues from the energy business in developing new waste collection services like Germany's Stadtwerke will help in solving the problems of the local community on the premise of population decline and post-COVID-19, and in giving back to the local public.

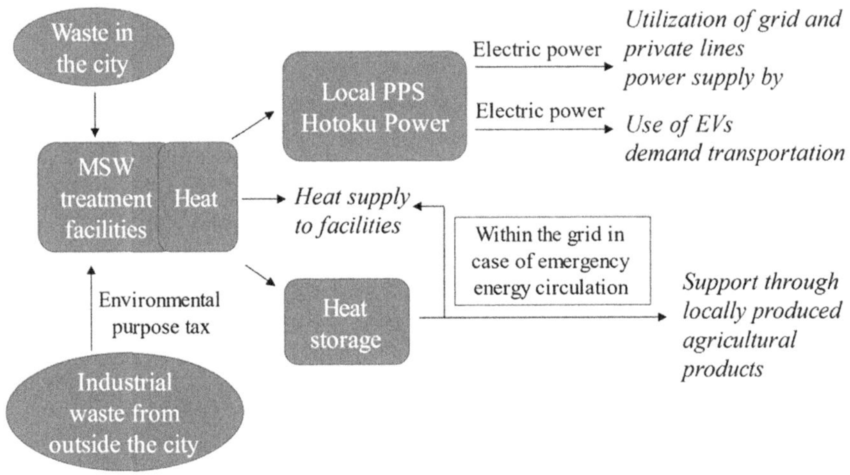

Fig. 8.9 Diagram showing the Kakegawa regional CES concept centered on MSW treatment facility. Prepared by the author based on interviews

In contrast to the method where citizens separate their own garbage, bring it to the garbage collection point on the day of collection, and the workers collect it in a unified manner, Kakegawa City has installed a smart trash can equipped with a weight sensor and an image recognition camera for each house in the smart community block. The city also designed a garbage disposal support service in which workers collect garbage from their homes according to the garbage collection rate. The people then can check the road and pavement status by using the drive coder during patrol. In addition, as further development for the business, the elderly monitoring service, which utilizes integrated management of electricity, water, and gas via smart meters, and the integrated home delivery management service at regional bases in the community block are expected to be introduced.

Such a Kakegawa regional CES concept shows the possibility of the Japanese version of Stadtwerke toward the realization of a sustainable MSW treatment system. In other words, in Japan, cities/municipalities must play a leading role in the administrative responsibilities stipulated by law to dispose of waste in the region so as not to cause any problems in the living environment. It is also the responsibility of the cities/municipalities to introduce cutting-edge technology while utilizing the expertise of private companies. Moreover, they also have to make sure that they take an approach to build a relationship of trust with citizens through the development of regional infrastructure and public services that are relevant to the daily and familiar aspects of the citizens' lives. These include garbage disposal support services for solving regional issues such as requiring elderly people to dispose of their waste, ensuring a hygienic living environment, and improving social acceptance by giving back to the local public. In Kakegawa, it is important to find a new PPP system by appropriately dividing the roles of the public and private sectors to build a sustainable MSW treatment system, and to work hard to realize a Japanese version of Stadtwerke.

8.6 Conclusion

In this paper, we discussed the sustainability of the MSW treatment system in Japan from two perspectives: PPP and prioritizing local public interest. In particular, through the case study of Kakegawa regional CES, the following issues and possibilities for improvement toward the establishment of a sustainable MSW treatment system are presented.

(1) Japanese version of Stadtwerke approach

In Japan, it is possible to develop a local PPS business by utilizing waste power generation when constructing a sustainable MSW treatment system. However, it is important to develop regional infrastructure to solve regional issues by using the profits earned from the energy business while taking up PPPs without being devoted to the electric power business. The aim is to build a Japanese version of the German Stadtwerke, wherein the reinvestment is for the provision of public services.

(2) Division of roles between public and private sectors in the PPP

In Japan, inefficiencies in MSW treatment facilities have become apparent due to the decline in waste discharge. In response, a mixed MSW and industrial waste combustion is being contemplated at private facilities after the public and private sectors (e.g., industrial waste producers) conclude a cooperation agreement. Such new public–private partnership methods can ensure a stable supply of waste. However, while private businesses are closely linked to the region, it is also important to build a regional trust relationship through PPPs. This is in addition to the involvement of the government that promotes the decision of the residents in terms of the location of the facilities and creates regional value by providing high-quality public services. In addition, to achieve long-term and stable operation, it is necessary to seek an appropriate balance between the risks and benefits of the public and private sectors through the establishment of SPCs.

(3) Maximizing local public interest and social acceptance

In building a sustainable MSW treatment system, it is necessary to grasp the concrete image of maximizing local public interest in cooperation with the local PPS, build a relationship of trust with the citizens, and seek improvement in social acceptance. This can be achieved with the development of regional infrastructure that is relevant to the daily and familiar aspects of the citizens' lives and by providing public services as starting points.

For future work, it is necessary to comprehensively analyze and compare the case studies of MSW treatment systems not only in Japan, China, and Europe, but also those in the United States and Southeast Asia.

Acknowledgements This work was supported by JST SPRING, Grant Number JPMJSP2128. We would like to express our gratitude to all related parties.

References

Agency for Natural Resources and Energy HP [Online]. https://www.enecho.meti.go.jp/category/electricity_and_gas/electric/electricity_liberalization/what/
Aioi city HP [Online]. https://www.city.aioi.lg.jp/soshiki/kankyo/sinbika.html
Berlin city HP [Online]. https://www.berlin.de/
BSR HP [Online]. https://www.bsr.de/
Bureau of Environment HP [Online]. https://www.kankyo.metro.tokyo.lg.jp/resource/general_waste/about.html
Chichibu S (2015) Kobelco eco-solutions technical report trade, of participation in "The 14th investigation group on overseas environmental business" sponsored by Japan Environmental Facilities Manufacturers Association (JEFMA) 11(2):37–41
Chichibu Power Supply HP [Online]. https://chichibu-pps.co.jp/
Daoei Kankyo Group HP [Online]. https://www.dinsgr.co.jp/news_no200/
European Commission HP [Online]. https://ec.europa.eu/eurostat/documents/2995521/7214320/822032016-APEN.pdf/eea3c8df-ce89-41e0-a958-5cc7290825c3
Fushan Circular Economy Trade and Industry Park HP [Online]. https://www.askci.com/xmal/20190624/1203301148258.shtml
Fukuyama Mirai Energy HP [Online]. https://fukuyama-miraienergy.co.jp/
Guangzhou city HP [Online]. http://www.gz.gov.cn/
Hamamatsu Power Supply HP [Online]. https://www.hamamatsu-e.co.jp/
Hata M (2022) The potential for public-private partnerships in waste management in municipalities with declining populations Collection of papers presented at the 43rd lecture meeting of the national conference on urban cleaning research and case studies
Inagaki K (2017) Efforts by local PPS in using power sources owned by regional governments [Online]. https://www.kankyo-business.jp/column/014251.php
Japan Environmental Sanitation Center (2017) Study group on the future of waste power generation second stage final report [Online]. https://www.jesc.or.jp/Portals/0/center/activity/saisei/H29gomisaishu2.pdf
Japan Environmental Sanitation Center (2020) Investigation and research to diffuse and promote the low carbon and CO2 reduction measures through waste treatment systems and commissioned feasibility analysis [Online]. https://www.env.go.jp/recycle/report/r3-08/co2.html
Kakegawa and Kikugawa Sanitation Facility Association (2021) Basic concept of waste treatment facility development [Online]. http://www.kankyoshigen-gallery.jp/info/pdf/20210330document01.pdf
Kakegawa City (2021a) About Kakegawa local PPS business [Online]. http://www.pref.shizuoka.jp/sangyou/sa-150/vpp/documents/r2_2siryou3.pdf
Kakegawa City (2021b) Electricity supply to public facilities [Online]. https://www.city.kakegawa.shizuoka.jp/fs/1/4/8/9/5/4/_/2021_01_tokushu2.pdf
Kakegawa City (2021c) Toward the realization of Kakegawa's version of Stadtberke [Online]. https://www.city.kakegawa.shizuoka.jp/fs/1/4/8/9/5/5/_/2021_02_tokushu3.pdf
Kakegawa city HP [Online]. https://www.city.kakegawa.shizuoka.jp/
Kakegawa Hotoku Power HP [Online]. https://kakegawa.de-power.co.jp/
Kakegawa kankyoshigen-gallery HP [Online]. http://www.kankyoshigen-gallery.jp/aboutus/index.html
Kikugawa city HP [Online]. https://www.city.kikugawa.shizuoka.jp/
Kitakyushu city HP [Online]. https://www.city.kitakyushu.lg.jp/
Kitakyushu Power HP [Online]. https://kitaqpw.com/
Kumamoto Prefecture HP [Online]. https://www.pref.kumamoto.jp/
Liu Y (2021) Reference of top-level design of risk allocation of PFI project in Japan to PPP legislation in China [Online]. https://www.sohu.com/a/466241385_480400
Marubeni Ina Power HP [Online]. https://www.ina-mirai-denki.com/
Mashiki town HP [Online]. https://www.town.mashiki.lg.jp/

Minamikawa H (2016) Progress in waste management in environmental health Collection of papers presented at the 11th Lecture Meeting of the MSW treatment facility Management
Ministry of Economy, Trade and Industry (2017) Report on the progress of general liberalization of the electricity retail market [Online]. https://www.meti.go.jp/shingikai/enecho/denryoku_gas/denryoku_gas/pdf/003_03_00.pdf
Ministry of the Environment (2019) Wide-area waste treatment and consolidation of waste treatment facilities to ensure sustainable and appropriate waste treatment [Online]. https://www.env.go.jp/hourei/11/000652.html
Ministry of the Environment (2014) History and current status of waste management in Japan [Online]. https://www.env.go.jp/recycle/circul/venous_industry/ja/history.pdf
Ministry of the Environment (2021a) Dioxin emissions from MSW treatment facilities [Online]. http://www.env.go.jp/recycle/dioxin/ippan/post_89.html
Ministry of the Environment (2021b) Concentration of dioxins in exhaust gas from MSW incineration facilities [Online]. http://www.env.go.jp/recycle/r2ippaisannpaikouhyou.pdf
Ministry of the Environment HP [Online]. https://www.env.go.jp/seisaku/list/kyoseiken/index.html
Ministry of the Environment Results of municipal solid waste disposal survey [Online]. https://www.env.go.jp/recycle/waste_tech/ippan/stats.html
Ministry of the Environment Progress in the total deregulation of electricity and gas retailing [Online]. https://www.meti.go.jp/shingikai/enecho/denryoku_gas/denryoku_gas/pdf/037_03_00.pdf
Narita Katori Energy HP [Online]. https://www.nk-energy.jp/
Niigata Swan Energy HP [Online]. https://niigata-se.co.jp/
Ogawa A, Onoda H (2020) Feasibility study of cooperation between waste incineration plants and sewage treatment plants. CEIS Conf Environ Inf Sci 34:305–310
Onoda H (2018) Advisable technological developments in the environmental and energy fields, based on policies aimed for circular and low-carbon societies Collection of papers presented at the 31st Lecture Meeting of the Environmental Engineering Association
Onoda H (2020) Smart approaches to waste management for post-COVID-19 smart cities in Japan. IET Smart Cities 2(2):89–94
Onoda H (2022) Future waste treatment systems for decarbonization—perspectives and responses required for facility development. INDUST 412:2–7
Osako M (2022) The state of the technology system in the field of resource recycling and waste toward carbon neutrality in 2050. J Environ Technol 186:17–21
Oxfordshire County HP [Online]. https://oxfordcity.co.uk/
Shinjuku city HP [Online]. https://www.city.shinjuku.lg.jp/seikatsu/file09_02_00001.html
Smart Energy Kumamoto HP [Online]. https://se-kumamoto.co.jp/
Takaoka M (2020) Possibility of integrated incineration treatment of waste in a RegionalCES [Online]. https://jsmcwm.or.jp/mswi/files/2020/05/jsmcwm-shoukyaku-20200528-all.pdf
Viridor HP [Online]. https://www.viridor.co.uk/energy/energy-recovery-facilities/ardley-erf/
Wuppertal Institut (2018) Status und Neugründungen von Stadtwerken. Deutschland und Japan im Vergleich. Inputpapier zum Projekt Capacity Building für dezentrale Akteure der Energieversorgung in Japan. Wup-pertal
Wuzhou city HP [Online]. http://www.wuzhou.gov.cn/
Wuzhou Circular Economy Industrial Park HP [Online]. http://www.gxcic.net/subject/ztbd/showgjgc.asp?ID=173519
Yoshidome D, Kikuchi R, Okanoya Y, Pandyaswargo AH, Onoda H (2021) Actual measurement and evaluation of the balance between electricity supply and demand in waste-treatment facilities and development of adjustment methods. Appl Sci 11(22):10747
Zhao S, Onoda H (2021) Issues and potentials of local PPS in realizing a Japanese version of Stadtwerke. JOCET 9(1):12–23

Chapter 9
Geotourism and the Effects Caused by Solid Waste in the Tourist Attraction of Geological Formations of Torre Torre – Huancayo, Peru

Jeffri Steve Quispealaya Marin, Yeminna Zelha Huari Sanabria, Paola Andrea Jeremias Espinoza, Renato Saul Nino Bravo Verde, and Nelida Tantavilca Martinez

Abstract Geotourism in the "Geological Formations of Torre Torre" would be affected by the presence of solid waste. The objective is to determine the effects caused by solid waste in geotourism. Three zones were divided into sectors: High Zone (ZA), Middle Zone (ZB) and Low Zone (ZC), identifying representative geotopes in order to propose a geotourism route and determine the effects of waste through critical points in relation to geotopes; through surveys, visitors made known their perception of the environmental and geological aspect. 12 geotopes and 34 critical points for waste were identified, mainly organic with 54.22%, according to the perspective of people, the ZA is the most attractive and affected by solid waste, in addition to 83.12% think that geotopes are affected by these. On the other hand, visitors are unaware of the geological aspects of the place and the classification of waste; Based on the location of the geotopes, access and recommendation of the people, the geotourism route was developed. In conclusion, the waste does not generate negative effects on the geological heritage or the perspective of the visitors on their

J. S. Quispealaya Marin (✉) · Y. Z. Huari Sanabria · N. Tantavilca Martinez
Universidad Continental, Escuela Académica Profesional de Ingeniería de Minas, Huancayo, Perú
e-mail: 76921821@continental.edu.pe

Y. Z. Huari Sanabria
e-mail: 72908111@continental.edu.pe

N. Tantavilca Martinez
e-mail: ntantavilca@continental.edu.pe

P. A. Jeremias Espinoza · R. S. N. Bravo Verde
Universidad Continental, Escuela Académica Profesional de Ingeniería Ambiental, Huancayo, Perú
e-mail: 75492007@continental.edu.pe

R. S. N. Bravo Verde
e-mail: 70345937@continental.edu.pe

J. Baeyens et al. (eds.), *Proceedings of 2022 4th International Conference on Environment Sciences and Renewable Energy*, Environmental Science and Engineering, https://doi.org/10.1007/978-981-19-9440-1_9

return; however, it is proposed to promote geotourism as an option for sustainable development.

Keywords Geotourism · Solid waste · Geotope

9.1 Introduction

Tourism is represented by the beauty of natural resources. Being geotourism as a current trend and with a sustainable approach based on geological heritage. The National Geographic Society defines geotourism as tourism that sustains or enhances the geographic character of a place: its environment, culture, aesthetics, and the well-being of its residents (Dowling 2022). To date, through UNESCO, 161 Geoparks have been established worldwide, in 44 countries and 4 Global Geoparks Regional Networks that work for the Global Geoparks Network (GGN) (UNESCO 2022).

In Peru, the Geological, Mining and Metallurgical Institute (INGEMMET) through the "Geological Heritage and Geotourism" program encourages education, conservation and sustainable use of resources, registers places of geological interest and above all promotes geotourism and creation of geoparks, inventorying 450 sites of geological interest in regions such as Lima, Ica, Pasco and Junín (Geological Mining Institute 2021). However, in our country, the progress of geotourism is limited due to the existence of disagreements between authorities and people. The department of Junín has different tourist attractions, of which we can highlight a natural site called "Geological Formations of Torre Torre", located in the province and district of Huancayo (Ministry of Foreign Trade and Tourism 2020). In addition, these formations are made up of silt, clay and gravel modelled by a topography of gentle to steep slopes, affected by the erosion of gullies and the channelling of streams from 3300 m above sea level, highlighting their sedimentary nature (Ayala 2016). For this reason, it is important to apply the concept of geotourism due to its geological characteristics that have not yet been exploited, due to inefficient tourist and environmental management and deterioration of the attractiveness.

While it is true, the issue of solid waste generation is increasing at an accelerated rate. In 2018, in the report What a Waste 2.0 of the World Bank, it is estimated that in the next 30 years, the amount of waste will increase by 70% worldwide and by 2030 the generation per capita per day in Latin America and the Caribbean range to 1.11 kg (Kaza et al. 2018). However, solid waste management is a universal problem, but also important in achieving the Sustainable Development Goals (SDGs). Tourism represents about 20% of the GDP of several countries, being the third most essential sector in the world economy (United Nations 2020). However, it will be the main key to achieving the 17 SDGs and reaping their benefits (World Tourism Organization 2021); and avoid the externalities generated in the environment that are little attended. All the activities carried out by the human being produce solid waste, being generated in greater quantity in tourist places, thus affecting the environment. An essential factor is that they must be kept clean, so as not to lose their tourist potential; however, there is no environmental awareness among visitors and the people who live around it. Since its declaration as a protected and intangible area of the "Natural Geological Zone and Tourist Attraction of Torre Torre" (Provincial Municipality of Huancayo 2016), the management has not been in accordance with its conservation objectives since, areas contaminated with solid waste are visualized in the route of the area.

Geotourism in research will represent alternative tourism, conscious and sensitive to the interaction of people with the environment. Seeking thus, a scenic content and the beauty of the natural environment through the identification of geotopes, the interaction of geology and the environment, synthesizing in a proposed geotourism route.

9.2 Materials and Methods

9.2.1 Description of Study Area

The tourist attraction of "Geological Formations of Torre Torre", a tourist resource located in the department of Junín, province and district of Huancayo located 2.4 km east of the city, over 3400 m above sea level. Naturally formed by the erosive action of rain and wind, giving rise to columns of sedimentary rocks approximately 30 m high, made up of clay in pyramidal and triangular shapes, among others (Ministry of Foreign Trade and Tourism 2020). Municipal Ordinance No. 321-MPH-CM indicates a total extension of 910,128.92 m2 (Provincial Municipality of Huancayo 2016).

However, using a GPS, the study area was delimited with an extension of 61,828.11m2 within the geological zone (Fig. 9.1). Likewise, for the evaluation of the geotopes, it was divided into 3 zones called High Zone (ZA), Middle Zone (ZB) and Low Zone (ZC) (Fig. 9.2).

9.2.2 Geoturism

Collection of Information on Geological Aspects. Surveys were used, collecting quantitative and qualitative data to know the visitors' perception of geological aspects. Therefore, 77 surveys were conducted on two dates: September 25 to 43 people and October 9 to 34 people, from 9:00 am to 3:00 pm, asking ten closed questions.

Geotope Identification. For the Colombian Geological Service, a geotope is a delimited spatial portion of the geosphere for being a place of interest and representative for its study and geological interpretation (Colombian Geological Service 2021). Therefore, it is important to keep them preserved and protected. The identification of points of geological interest was carried out through observation throughout its journey, in addition to consultation with the inhabitants. Afterwards, a Garmin eTrex

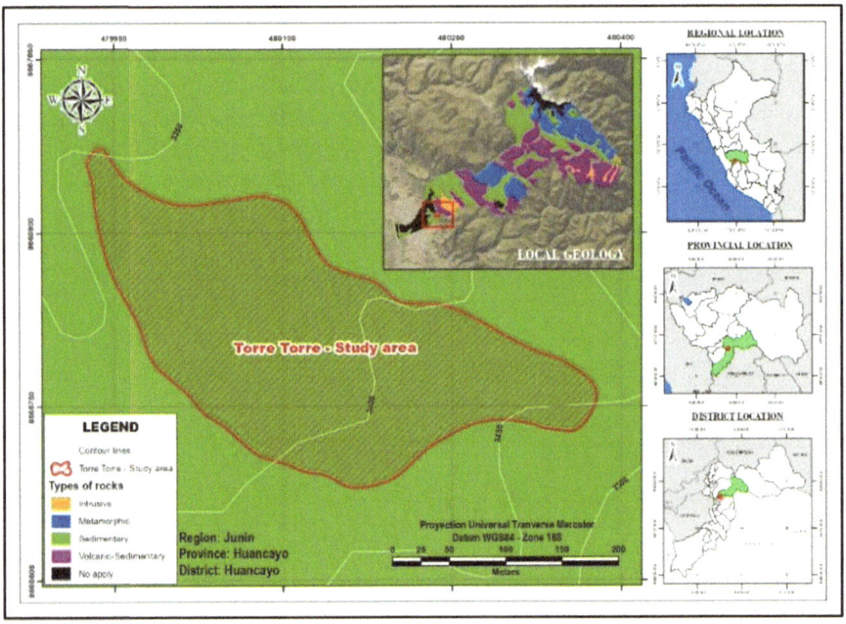

Fig. 9.1 Location, delimitation of the study area

Fig. 9.2 Sectorization of the zones (ZA, ZB and ZC)

10 model GPS was used to determine the location of the points by means of the coordinates in the three zones and a map was created in the ArcGIS v10.4.1 software. Likewise, a photographic camera was used to describe the geotopes.

Geotourist Route Proposal. Geotourism is focused on the improvement of geological features, historical heritage, and stability of people (UNESCO 2021). For the proposal of the geotourism route, a survey of the roads with GPS was developed considering the accesses and location of the geotopes to prepare a map.

9.2.3 *Solid Waste*

Identification of Critical Points Affected by Solid Waste. It is important to recognize the areas altered by the accumulation of solid waste, since they cause negative environmental impacts in the alteration of the soil and deterioration of the landscape perspective, putting at risk the use of the tourist resource. For this reason, it was essential to identify the critical points for solid waste taken with a GPS and taking into account the amount and type of waste.

Analysis on the Classification of Solid Waste. The amount of solid waste generated is variable within an attraction, since it depends on the number of visitors per day and its composition. Therefore, the analysis consisted of determining the amount generated per capita, in such a way that the visitors during their tour collected their waste in plastic bags that had previously been given by the research team. Later, at the end of their tour, the visitors handed over their collected waste to the team for later weighing, using an electronic scale. Then all the waste was classified according to the Methodological Guide for the development of the Characterization Study of Municipal Solid Waste (EC-RSM) (Ministry of the Environment Peru 2018), in order to weigh them on the scale to know the total amount of waste generated by each type.

9.2.4 *Evaluation of the Effects of Solid Waste in Geotourism*

Solid waste can negatively influence the use of tourism resources. For this reason, it was essential to use a qualitative technique (surveys) carried out on all visitors after carrying out their tour of the tourist resource, which made it easier to know their previous concepts and perspectives to identify and interpret the problems and provide solutions that helped reach the objective. They were raised with five open and closed questions.

With the information collected, the statistical base was obtained, which allowed ratifying the situation of the problem with respect to the effects of solid waste on geotourism potential. Likewise, the NTP 900.058.2019 "Color code for the storage of solid waste" (INACAL 2019) was taken into account.

9.3 Results

9.3.1 *Geotourism*

Description on the Geological Perception of Visitors. The result was that 22.08% know the geological aspects of the attraction and 77.92% aren't know. With the same result regarding the knowledge of geotourism. 54.55% do not know the type of material that makes up the structures (Fig. 9.3).

16.44% of those surveyed visited this tourist place through social networks, the main being Facebook, 10.39% by tourist companies, 1.30% by radio and television, family recommendations 33.77% and other media 19.48%, likewise, 20.78% already knew that place. All the visitors found the entire tourist route very interesting, in addition to its geological formations. However, the aspects that influence the return of people are solid waste in its highest percentage with 64.94%, 43.28% the deterioration of natural areas and 28.57% citizen insecurity (Fig. 9.4).

On the other hand, 76.62% of the visitors consider that the upper zone is of greater visual attraction due to the geotopes present, followed by 42.86% in the middle zone and 18.18% in the lower zone (Fig. 9.5).

According to the perspective of the visitors, 83.12% consider that the state of the geological zones is altered by solid waste (Fig. 9.6).

Geotopes in the Tourist Attraction. In the tourist attraction, a total of 12 representative geotopes perceived by the research team, tourist guides and local people were identified. Located 3 in the ZC (Fig. 9.7 and Table 9.1), 4 in the ZB (Fig. 9.8 and Table 9.2) and 5 geotopes in the AZ (Fig. 9.9 and Table 9.3). On the other hand, the description of each of these was based on morphology, which allowed them to be named uniquely.

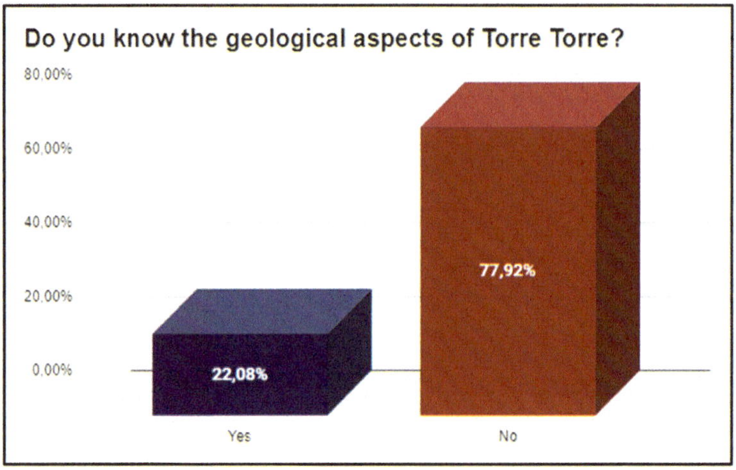

Fig. 9.3 Knowledge of visitors regarding geological aspects

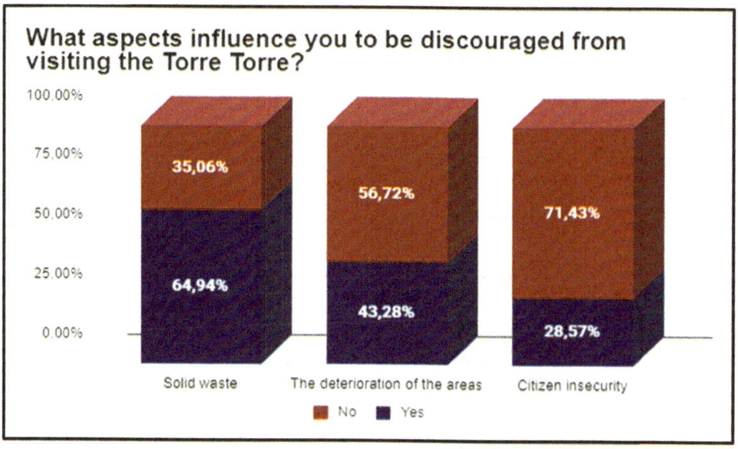

Fig. 9.4 Influential aspects for the return to the tourist attraction

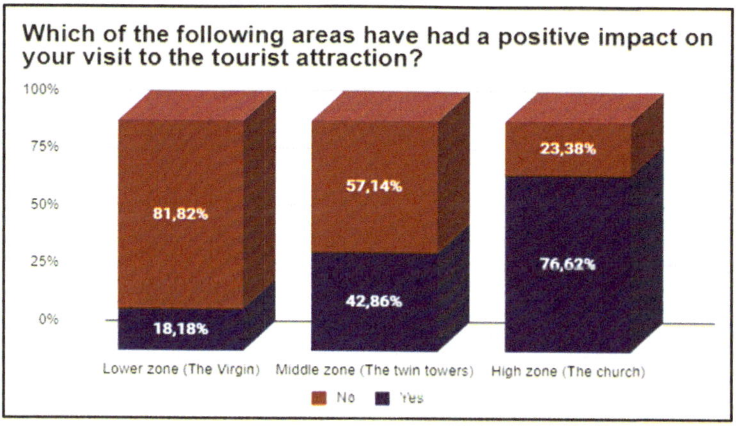

Fig. 9.5 Tourist attraction areas

Geotourism Route

The Geotourism Route is an Access Road to a Geologically Tourist Area, for This Reason a Possible Scenario was Proposed Through the Identified Geotopes.

Possible Scenario. The 12 geotopes were the fundamental basis for the design of the proposed geotourism route. This route has a length of 607.4274 m. throughout the tour of the tourist circuit.

The main entrance represents the beginning with a green flag on the map, located at 479,971.24 m to the East and 8,666,820.66 m to the North, the access road has an approximate width of 0.80 m in the first sections, at reaching the middle zone

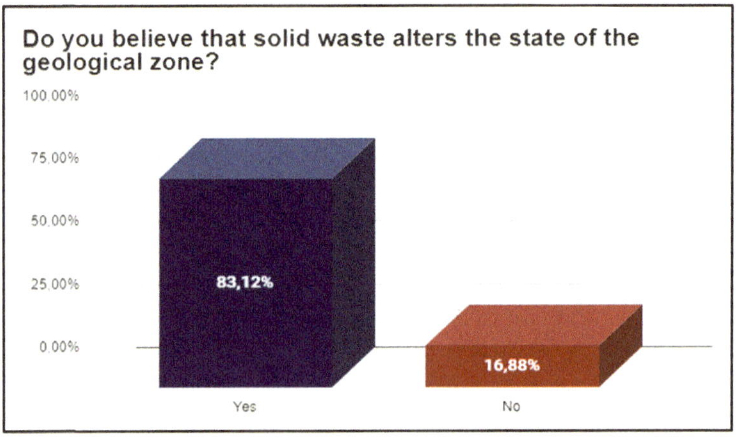

Fig. 9.6 Alteration of solid waste on the state of geological zones

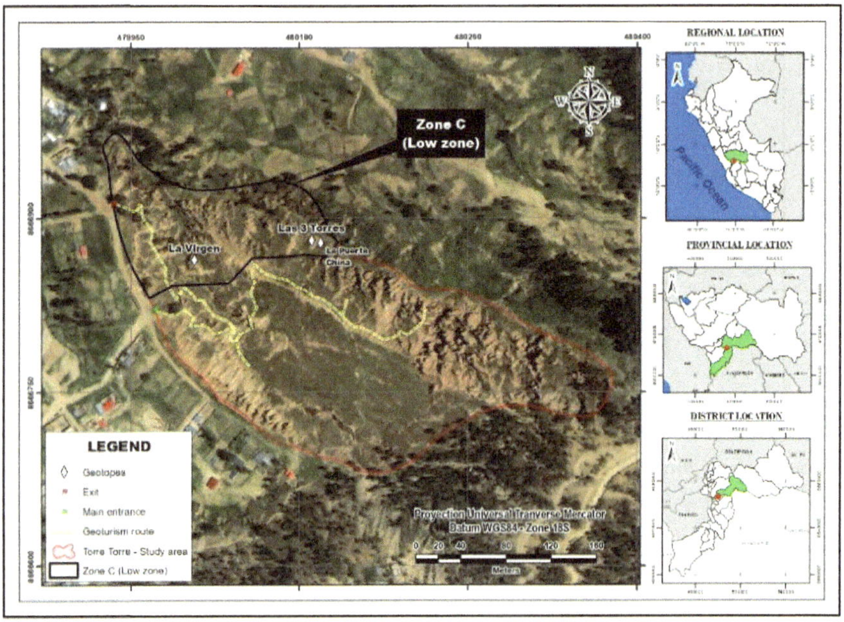

Fig. 9.7 Geotope map of the lower area

is the geotope called "Las Torres Gemelas", continuing the route "La Pirámide" is located near a small stream and a few meters away you can see "La Cabeza del Perro". To enter the lower area, you return to the geotope of "Las Torres Gemelas" passing through a hill, to visualize "La Puerta China" and "Las 3 Torres". Next, the route is followed where you can see the geotope "La Cabeza del Inca" on the left

Table 9.1 Morphological description of the geotopes in the lower zone

Photography	Description
	The geotope is at 480,007 m. to the East and 8,666,864 m. to the North and is located at 172.95 m. of income. Named as "La Virgen". The virgin is seen standing holding a wand wearing a crown and a long dress
	The geotope called the "Las 3 Torres" located at 480,111 m. to the East and 8,666,881 m. to the North and from the main entrance it is 262.55 m., where three staggered pillars can be seen. Also, there are windows at the ends and a corridor connecting the 3 towers in the middle
	"La Puerta China" is a geotope in the shape of an arch, it is part of the traditional Chinese architecture that is also called Paifang, located at 480,119 m. to the East and 8,666,879 m. to the North, with a distance of 265.62 m. from the main entrance

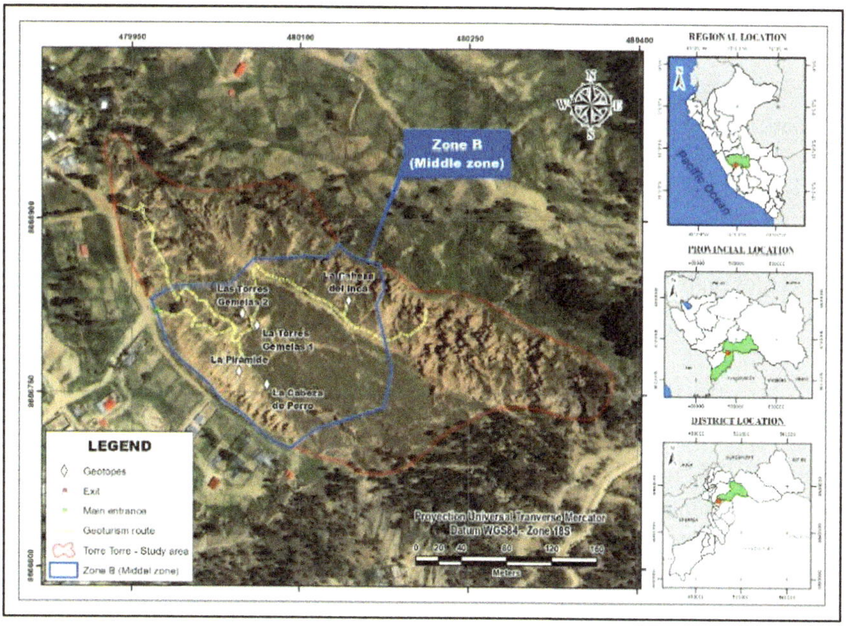

Fig. 9.8 Geotope map of the middle zone

Table 9.2 Morphological description of the geotopes in the middle zone

Photography	Description
	"Las Torres Gemelas" is made up of two columns of similar size with a distance of 16.4 m from each other, the first located at 134.59 m. from the entrance, having as coordinates to the East 480,062 m. and to the North at 86,668,808 m. The second column is located at 145.03 m. from the entrance with coordinates to the East of 480,049 m. and to the North 8666818 m
	"La Pirámide" is located at 137.4 m. from the entrance with coordinates to the East of 480,046 m. and to the North 8,666,768 m. It represents a set of pillars forming a pyramid in the center of the image. A rectangular door located in the lower half is also displayed
	The geotope called "La Cabeza de Perro" is at 480,071 m. to the East and 8,666,757 m. to the north. located at 153.21 m. from the starting point. The profile of a dog's head is observed, distinguishing features such as: the nose, the snout and an ear
	"La Cabeza del Inca" is located at 297.09 m. from the entrance, it is at 480,143 m. to the East and 8,666,830 m. to the North, it has a particular shape of an Inca wearing a traditional crown and circular earrings

side and, in front of it, "La Bota de Duende" and "Los Enamorados". Approximately 40 m. higher up are "La Capilla" and "El Puente"; As a final geotope, in the upper area is "El Caballo de Ajedrez" which is considered as a turning point to return to "Las Torres Gemelas". The tour ends in the lower area where you can see the right side of the route to "La Virgen". The exit is represented with a red flag located at 479,933.83 m. East and 8,666,912.12 m. North (Fig. 9.10).

9.3.2 *Solid Waste*

Critical Points. It was possible to identify 34 critical points affected by solid waste in the three delimited areas, located in the vicinity of the geotourism route and geotopes (Fig. 9.11). Waste such as packaging, plastic, paper, cardboard, etc. was found. Also, due to the situation of the COVID-19 pandemic, surgical masks were found.

The critical points are distributed, identifying 8 points in the High Zone (ZA), 14 in the Middle Zone (ZB) and 12 in the Low Zone (ZC). These points affected by the accumulation of waste are located near the geotourism route and geotopes such as "La Virgen" in the Lower Zone, "Las Torres Gemelas" and "La Pirámide" in the Middle Zone and "La Capilla" in the Upper Zone. and "The Bridge". While

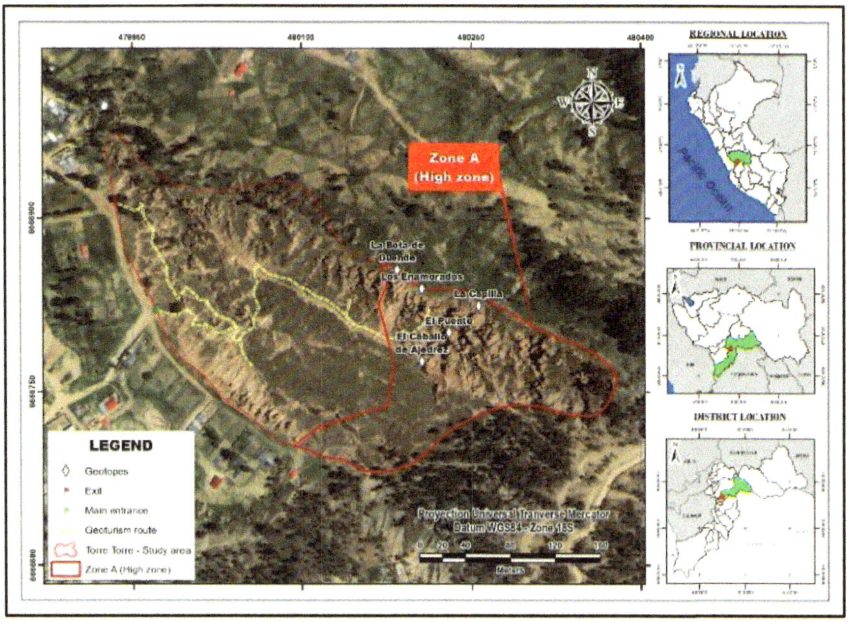

Fig. 9.9 Geotope map of the upper area

others are in free areas that are taken as rest by visitors in the ZA and ZB. Those that are not found in the areas mentioned above, are located in depths and between the formations.

Solid Waste Quantification. The waste collected by visitors was mainly of the type of usable (organic and inorganic), non-usable and dangerous waste (Fig. 9.12).

Where the largest amount of solid waste found on the two dates, are usable organic waste with 61.77 and 46.67% of the total generated on the first and second day respectively. On the other hand, what was collected in the inorganic waste is of the plastic type with 23.55% on the first day and on the second day with 38.89%, specifically PET (Polyethylene Terephthalate) and HDPE (High Polyethylene). Density), followed by tetra brik, metal and cardboard in a lower percentage on both days. Meanwhile, no waste such as paper, glass and textiles were found within said classification. Regarding non-usable waste, it was 4.59 and 2.78% in the two days, highlighting waste such as single-use plastic bags, Styrofoam and product wrappers. In the case of hazardous waste, 1.53% were obtained on the first day, compared to the second day, when no waste of this type was found.

Table 9.3 Morphological description of the geotopes in the upper zone

Photography	Description
	The geotope is located at 373.61 m. from the entrance with coordinates for the East is 480206 m. and 8,666,776 m. To the north, called "El Caballo de Ajedrez", there is mainly a horse that is a fundamental piece of the game
	The geotope is located at 480,230 m. to the East and 8,666,801 m. to the North, located at 385.58 m. From the main entrance along the geotouristic route, known as "El Puente", a bridge can be seen that, due to its particular shape, resembles the popular Golden Gate Bridge
	The point of geological interest called "La Capilla", has the representation of a chapel with the dome in the deepest part and the apse. It is at 480,256 m. to the East and 8,666,824 m. to the North, it is located at 422.21 m. from the main entrance
	"Los Enamorados" is at 480,206 m to the East and 8,666,839 m to the North, it is located at 396.60 m. of the main income. You can see two towers face to face representing two people staring at each other with love
	"La Bota de Duende" is a geotope located at 426.11 m. from the starting point, it is 480185 m. to the East and 866,685 m. to the North, it has a very peculiar shape of an elf boot, managing to see the instep at the bottom

9.3.3 *Effects on Geotourism in Relation to Solid Waste*

The results of the surveys are presented according to their knowledge and perception after having visited the tourist resource, as well as the analysis regarding the effects of solid waste and geotourism related to each other. The basic knowledge of the visitors about solid waste allowed us to identify that 100% know the concept of environmental pollution, while 85.12% understand what solid waste is. For the analysis of the types of solid waste, we rely on the NTP 900.058.2019, being four types of waste. 55.84% recognize usable solid waste, 59.74% unusable, 62.34% organic and 51.95% hazardous waste (Fig. 9.13).

Regarding the effects of solid waste on geotourism, 89.61% consider that there is an impact on the landscape and fauna; 85.71% believe that there is an affectation to the soil, 84.42% visualize an affectation to the flora, while 76.62% regarding the affectation of the geology (Fig. 9.14).

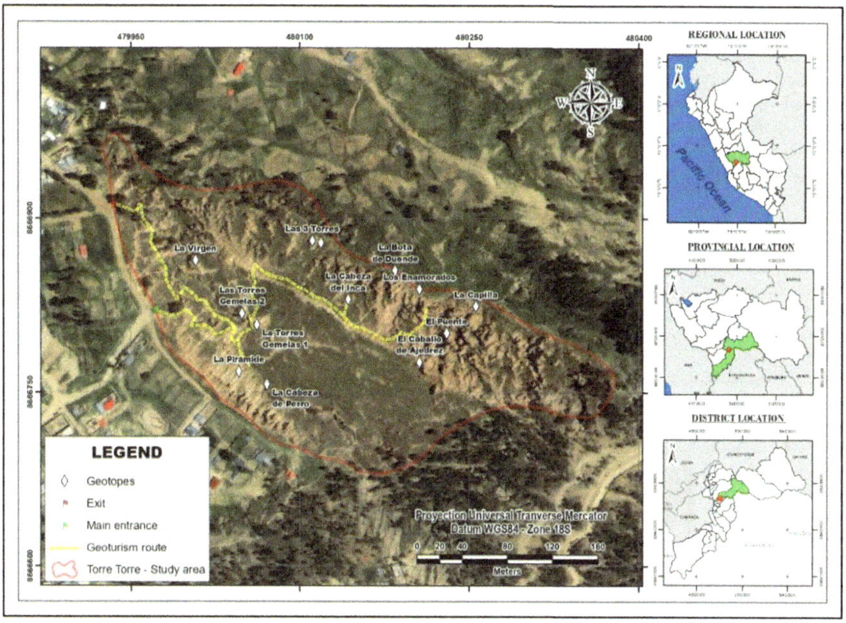

Fig. 9.10 Map of the proposed geotourism route

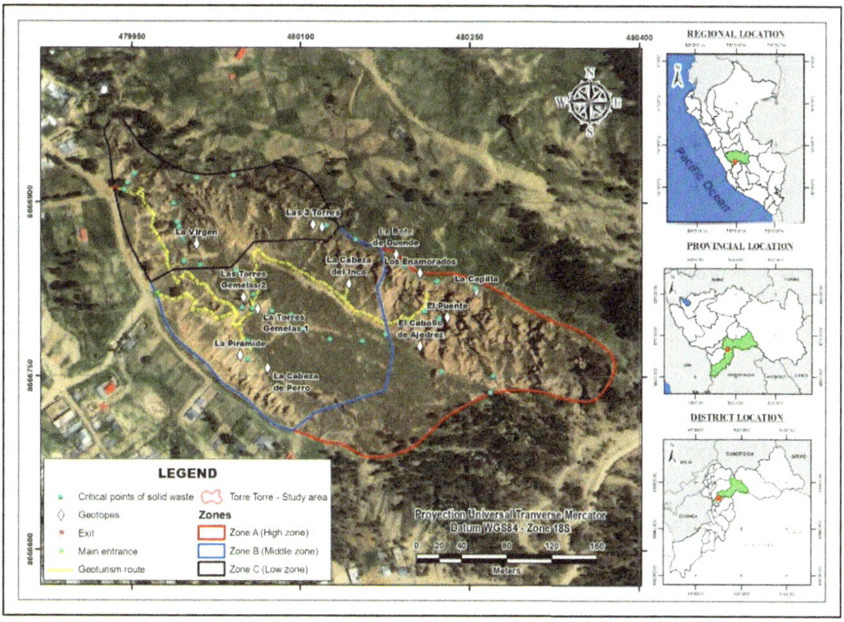

Fig. 9.11 Map of critical points of solid waste according to sectorization

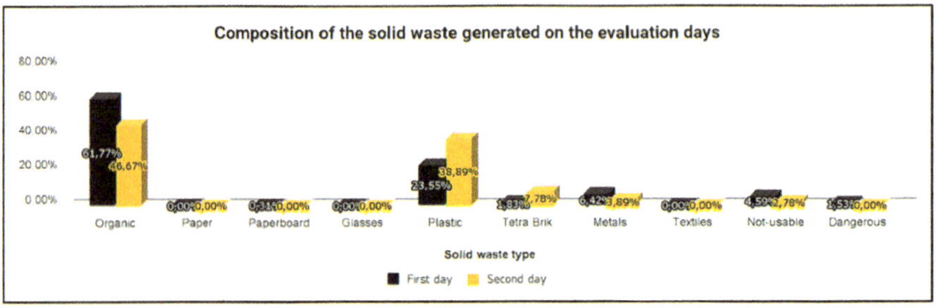

Fig. 9.12 Quantification of solid waste according to its classification

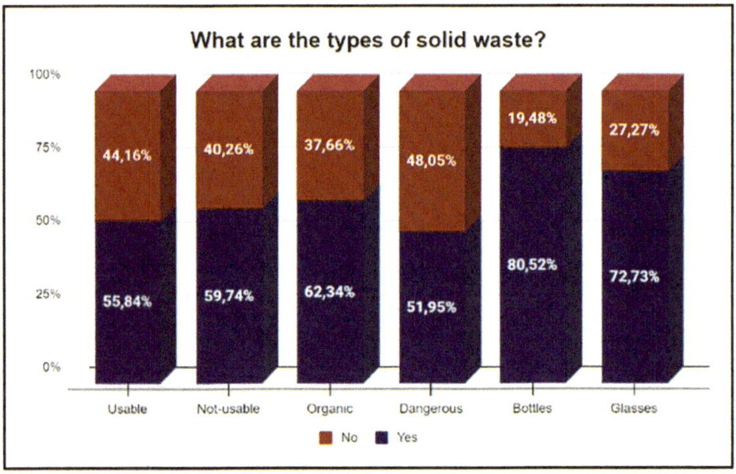

Fig. 9.13 Recognition of the types of solid waste by visitors

The perception of visitors regarding the affectation of the three zones and solid waste (Fig. 9.15) considers that the most affected is the high zone (ZA) with 61.04%; followed by 50.65% in the middle zone (ZB) and 29.87% in the low zone (ZC).

9.4 Discussion of Results

The research aims to project a geotourism in the attraction, through the relationship and comparison of results obtained in the field and the perception of the people who visited the place. According to the perspective of each visitor, a greater number of points of geological interest would be considered, so the most representative were determined in the study. However, according to the data taken in the field, the largest number of geotopes is found in the upper zone, compared to the critical points present in less quantity, due to the fact that visitors do not reach said zone due to its sloping

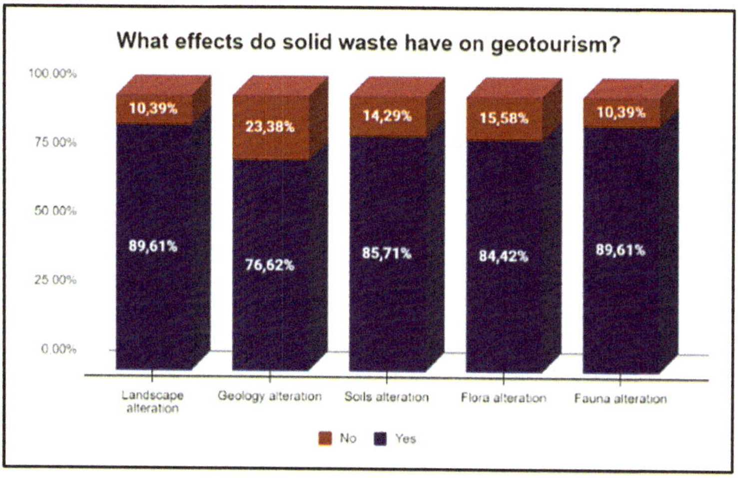

Fig. 9.14 Possible effects on solid waste

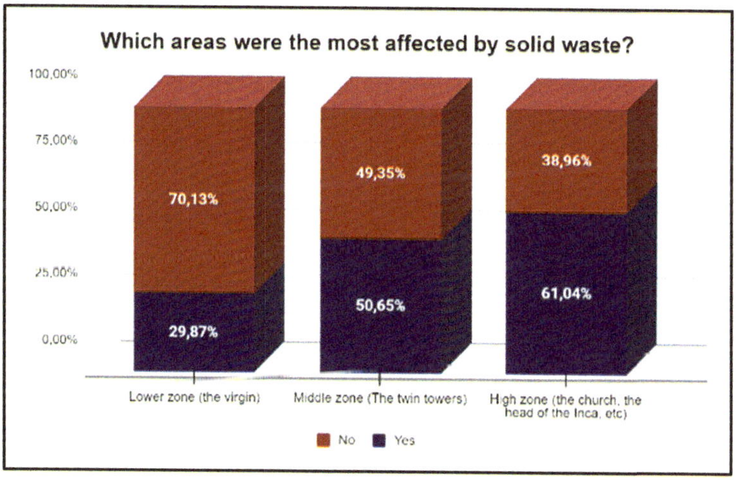

Fig. 9.15 Areas affected by solid waste

topography and its difficult access, why people do not dispose of their solid waste. Being the middle zone with the greatest number of critical points, since it is the intersection between the other two (ZA and ZC).

The perspective of the people shows that the upper zone is of greater attraction due to the number of geotopes, however it is the zone most affected by solid waste, due to the extension of each zone, since the visualization of the visitors could cover a greater compared to the lower area, where access roads are fewer. On the other hand, they perceive that solid waste alters geological aspects, but it is of less importance compared to landscape and fauna, because 77.92% do not know geological aspects. In addition, 64.94% of the visitors would not return due to the presence of solid waste in the study area, although as a whole they consider that they would return to the tourist resource due to the attraction of the points of interest and the landscape it presents.

9.5 Conclusions

The tourist attraction of Torre Torre Geological Formations is known for its geological nature; however, visitors are unaware of the geological aspects and type of material. In the field, 12 geotopes were identified that are mostly affected by solid waste and, according to the perspective of the visitors, they consider it a fundamental aspect for which they would not return. In view of the analysis, it is proposed to implement viewpoints towards the points of geological interest through the geotourism route.

According to the number of critical solid waste points distributed in the zones (ZA, ZB and ZC) and the appreciation of the people who visit said place, the need to implement differentiated containers in said zones with their respective signage was seen. Therefore, it is essential to carry out training on geotourism aimed at visitors, residents and authorities, so it should be implemented and promoted regarding geological aspects and solid waste management.

References

Ayala L (2016) Sedimentology of the Quaternary in Torre Torre

Colombian Geological Service: Geological and Paleontological Heritage: Defi-nitions (2021)

Dowling R (2022) Geotourism

Geological Mining Institute: Ingemmet highlights geotourism destinations on the Latin American and Caribbean Geotourism Day (2021)

Kaza S, Yao LC, Bhada-Tata P, Van Woerden F (2018) What a Waste 2.0: A Global Snapshot of Solid, Waste Management to 2050

Ministry of Foreign Trade and Tourism: Regional Strategic Plan for Tourism Junín, pp 2020–2025 (2020)

Ministry of the Environment Peru: Methodological Guide for the development of the Municipal Solid Waste Characterization Study (EC-RSM), pp 70 (2018)

Provincial Municipality of Huancayo: OM-MPH-CM No. 321 "Ordinance that declares the Natural Geological Zone and Tourist Attraction of Torre Torre as a Protected and Intangible Area" (2016)

Standardization Directorate – INACAL 2019. NTP 900.058:2019 WASTE MANAGEMENT. Color code for the storage of solid waste, pp 14 (2019)

UNESCO (2022). https://globalgeoparksnetwork.org/?page_id=4209, last accessed 4/05/2022

UNESCO Office in Mexico: Memory of the international meeting: Geoparks, Sustainable Tourism and Local Development, pp 227 (2021)

United Nations: Policy brief: COVID-19 and the transformation of tourism (2020)

World Tourism Organization: Tourism in the 2030 Agenda (2021)

Chapter 10
Utilization of New Fly Ash Type from Selective Catalytic Reduction (SCR) Process as an Additive in Portland Cement

T. Suwan, T. Jongwijak, P. Jitsangiam, C. Buachart, B. Charatpangoon, and K. Jitpairod

Abstract Fly ash is one of the by-products of coal-fired power plants for electricity generation. In Thailand, the largest plant is Mae Moh power station in Lampang province, which generates fly ash of approximately 6,000–8,000 tons per day. Those fly ashes can be used in the cement and concrete industry as an additive. However, the new installation of nitrogen gas (NO_x) treatment, called the "Selective Catalytic Reduction system (SCR)", could possibly leave some of the excess ammonia in the collected fly ash. The main aims of this study are to investigate the characteristics of the SCR-fly ash compared to the typical high calcium fly ash (HC-fly ash). Moreover, the evaluation of the properties of OPC-fly ash blended cement pastes using those two types of fly ash were also carried out. The preliminary results show that the SCR-fly ash can be used as an additive in cement and concrete, but, it should be rested for over a month to release the contaminated ammonia. The outcomes could be a source of information on utilizing this new type of SCR-fly ash and could also be an alternative approach for recycling industrial by-products.

Keywords Additives · Fly ash · Portland cement · Selective catalytic reduction process

T. Suwan (✉) · T. Jongwijak · C. Buachart · B. Charatpangoon · K. Jitpairod
Department of Civil Engineering, Faculty of Engineering, Chiang Mai University, Chiang Mai, Thailand
e-mail: teewara.s@cmu.ac.th

T. Suwan
Center of Excellence in Natural Disaster Management, Chiang Mai University, Chiang Mai, Thailand

P. Jitsangiam
Chiang Mai University Advanced Railway Civil and Foundation Engineering Center (CMU-RailCFC), Chiang Mai University, Chiang Mai, Thailand

J. Baeyens et al. (eds.), *Proceedings of 2022 4th International Conference on Environment Sciences and Renewable Energy*, Environmental Science and Engineering, https://doi.org/10.1007/978-981-19-9440-1_10

10.1 Introduction

Power generation in Thailand nowadays is still based on Coal-fired power plants. Fly ash is one of the by-products of the burning of coals which is trapped by an electrostatic precipitator before releasing to the atmosphere. Mae Moh power plant, the largest coal-fired power plant in Thailand, has a total electric power production of over 2,200 MW and generates fly ash of approximately 6,000–8,000 tons per day (EGAT 2020). In the initial stage, fly ash was defined as a waste that was dumped for the landfill. Many problems have occurred, for example, the contamination of toxins in the water and land, the spread of tiny fly ash particles to the nearby communities, or costly management budgets. However, in the past couple of decades, the usage of fly ash has been studied and developed. Fly ash is now effectively used in the cement and concrete industry as an additive in Portland cement (OPC) along with the ASTM C618 standard. It can reduce the amount of costly OPC consumption with some advantages of the concrete workability and the long-term strength improvements (Suwan et al. 2020). Fly ash can also be used to produce other alternative cementitious materials, such as alkaline activated cement and geopolymer cement (Maichin et al. 2020; Jitsangiam et al. 2021; Wattanachai and Suwan 2017; Bualuang et al. 2021). As aforementioned, all fly ash is thus entirely sold out to the cement and concrete manufacturers.

Coal-based power plants always come up with some pollutions such as the oxides of sulfur (SO_X), nitrogen (NO_X), or small particle matters. Therefore, specific treatment processes are installed to eliminate or reduce those amounts of the harmful toxins. The establishment of new power generators Units 4–7 (MMRP1) came with a new NO_X treatment system, known as the "Selective Catalytic Reduction system (SCR)". More efficiency on NO_X control for SCR technologies is achieved by injecting aqua ammonia or another reductant into a furnace or flue gas to convert NO_X to N_2. From the process of ammonia injection, there is a chance of unreacted ammonia escaping from the flue gas, yielding ammonia-contaminated fly ash called "SCR-fly ash" (Lu et al. 2020). With possible ammonia contamination, a massive amount of SCR-fly ash is, therefore, less confident to be used in the cement and concrete industry and kept unused in the field.

The main aims of this study are to investigate the characteristics of the SCR-fly ash in comparison with typical high calcium fly ash (HC-fly ash). Moreover, the evaluation of the properties of OPC-fly ash blended cement pastes using those two types of fly ash were also carried out. The outcomes could be a source of information on utilizing this new type of SCR-fly ash and could also be a preliminary result for further study.

Table 10.1 Chemical composition of OPC by XRF analysis

Materials	Al_2O_3	SiO_2	SO_3	K_2O	CaO	TiO_2	MnO	Fe_2O_3	SrO
OPC	2.46	11.93	4.89	1.03	74.96	0.41	–	4.06	–
HC-fly ash	8.61	17.99	7.71	2.55	31.73	0.49	0.22	30.31	0.29
SCR-fly ash	10.67	22.75	6.09	2.72	29.23	0.71	0.23	27.17	0.26

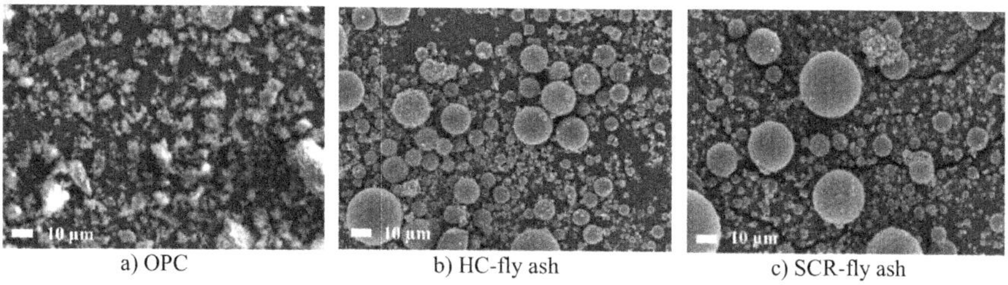

Fig. 10.1 Images of OPC, high calcium fly ash, and SCR fly ash from SEM

10.2 Materials

Portland Cement (OPC), SCG Type 1, complying with ASTM C150 and Thai Industrial Standard (TIS) 15-1 had a specific gravity of 3.15. Fly ashes (FA) were obtained from the Mae Moh coal-fired power plant, Lampang, Thailand. Fly ash from the Selective Catalytic Reduction system (SCR-fly ash) had a specific gravity of 2.58, while the average particle size and specific surface area were 39.05 μm and 0.516 m^2/g, respectively.

A typical high calcium fly ash (HC-fly ash) had a specific gravity of 2.85, while the average particle size and specific surface area were 14.46 μm and 0.879 m^2/g, respectively. Their chemical compositions by X-ray fluorescence (XRF) analysis are as shown in Table 10.1. Figure 10.1 shows the micrographs of particle's size and shape of OPC, HC-fly ash, and SCR-fly ash with 500 times magnification by the Scanning Electron Microscope (SEM).

10.3 Mixture Designations and Analytical Methods

10.3.1 Mixture Designations

Two types of fly ashes, HC and SCR, were prepared. The SCR-fly ash, in this study, was collected from the plant and kept dry-cool for approximately 200 days before testing. Each fly ash was initially dry-mixed with OPC for 2 min in a mixer as a cement binder. The proportions of OPC-to-fly ash (OPC:FA) were varied from 100:0, 80:20,

Table 10.2 Details of mixture proportions of fly ash-blended cement

FA type	w/c ratio	Mixtures	OPC:FA	Evaluation
SCR fly ash	0.35	SCR0	100:00	Setting time
		SCR20	80:20	Flowability
		SCR40	60:40	Compressive strength
		SCR60	40:60	Microstructure
HC fly ash	0.35	H.Ca0	100:00	Setting time
		H.Ca20	80:20	Flowability
		H.Ca40	60:40	Compressive strength
		H.Ca60	40:60	Microstructure

60:40, and 40:60, respectively. Then, the specific amount of water with a water-to-cement (w/c) ratio of 0.35 was added to the mixer. After well-mixing, freshly blended cement paste was neatly cast in the prism molds and wrapped with a plastic sheet to prevent moisture loss for 24 h. Next, after demolding, the samples were wrapped with the plastic sheet again and kept at room temperature until reaching the testing age of 3, 28, and 90 days. Table 10.2 presents the details of mixture proportions.

10.3.2 Analytical Techniques

A KCl Extraction technique was used to determine the concentration of ammonium residue (mg/kg) in the SCR-fly ash. The setting time of the paste was carried out by a Vicat apparatus (ASTM C191), while a flowability was done using a flow table test, according to ASTM C1437. The compressive strength test was carried out with a 40 × 40 × 40 mm. Specimen following an EN 196-1 standard. The specimens were de-molded 24 h after casting, then wrapped with a plastic sheet and left in the ambient conditions until reaching their ages of testing. The strength test was operated by a 250 kN Control universal testing machine (UTM). Scanning Electron Microscope (SEM), JEOL JSM-5910LV with 30 kV vacuum mode, was used to observe the micrographs and visual changes of the samples.

10.4 Results and Discussions

10.4.1 Ammonia Contamination in the SCR-Fly Ash

The ammonia content was measured after SCR-fly ash collection at the passing 10, 20, 30, and 200 days (see the bottom line-Mae Moh). The amount of ammonia contamination decreased by the time from 12.42, 6.93, 6.17, and to 2.17 mg/kg,

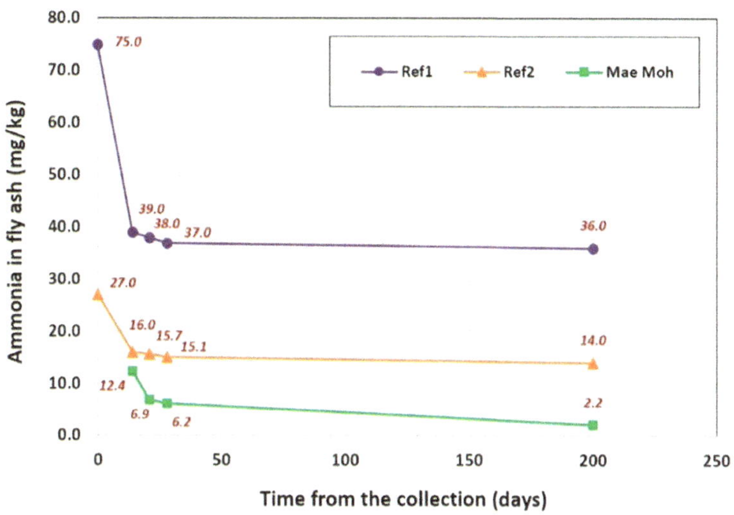

Fig. 10.2 Ammonia in SCR-fly ash at the different times from the collection

respectively (Fig. 10.2). Around a haft of ammonia disappeared in the first 20 days after exposure to the atmosphere. Similar trend was observed in the previous studies of Michalik et al. (2019) that the ammonia content in fly ash was dramatically decreased in the first couple of weeks (Michalik et al. 2019). Another researcher (Fisher et al. 1997) also reported that the amount of excess ammonia in the fly ash can be in the ranges of 50–300 mg/kg (Fisher et al. 1997; Larrimore 2002). Those observations had much more ammonia content than Mae Moh SCR-fly ash (see Ref1 and Ref2) (Michalik et al. 2019). Nevertheless, please be noted that there was no ammonia detected in the HC-fly ash in this study.

10.4.2 Setting Time and Flowability of the Pastes

Figure 10.3 shows the setting time of OPC-fly ash blended cement pastes. Both SCR- and HC-fly ashes in the cement pastes had a longer setting time than the typical OPC as fly ash slowed down the reaction in the early stage. More addition of fly ash, therefore, directly extended their setting behaviour. However, the setting time of SCR-fly ash seemed to be longer than that of HC-fly ash because of its larger particle sizes, leading to gain lower hydration reaction. The smaller particle size with the high surface area of HC-fly ash provided faster water absorption and reaction than the SCR type, leading to less flowability. Nevertheless, it was found that too much presence of fly ash led to the over-scale measurement of the flow table testing (Fig. 10.4).

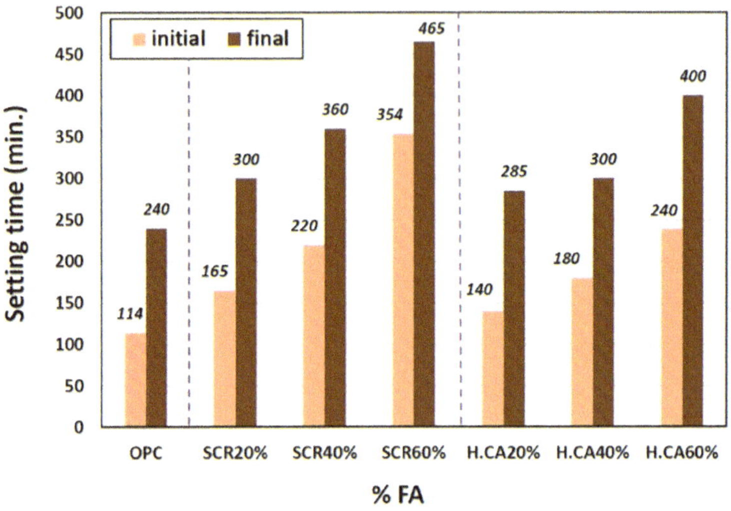

Fig. 10.3 Setting time of OPC-fly ash blended cement paste

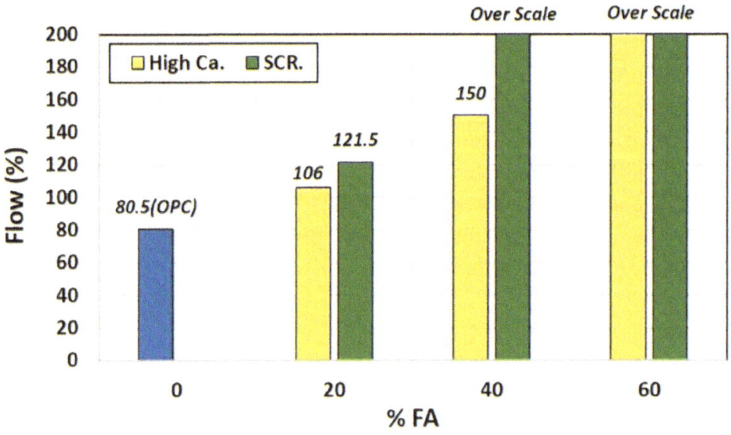

Fig. 10.4 Flowability of OPC-fly ash blended cement paste

10.4.3 Compressive Strength and Microstructures of the Blended Cement Pastes

The addition of fly ash to OPC has many advantages. For example, it reduces the concrete cost and lowers the heat of hydration, while increasing workability and long-term strength. However, according to ACI 318M-99, the maximum additional fly ash in the concrete shall not exceed 25% by the weight of cementitious materials (ACI Committee 318 1999). The results in Fig. 10.5 clearly define that the addition of

20% of fly ash provided the highest compressive strength than that of 40% and 60%, respectively. In contrast, the compressive strength of long-term age (90 days) was evidently increased due to an additional of the secondary calcium silicate hydrate (C-S-H) formation from the pozzolanic reaction between $Ca(OH)_2$ and fly ash (Suwan et al. 2020).

At all percentages of fly ash addition, the HC-fly ash mixtures achieved slightly higher strength than that of SCR-fly ash mixtures. Apart from the larger particle size of SCR-fly ash than HC-fly ash, the presence of ammonia (as ammonium sulphate; $(NH_4)_2SO_4$) was reported to react with calcium hydroxide ($Ca(OH)_2$), forming unwanted gypsum (see Eq. 10.1) (Alexander et al. 2013). This gypsum can quickly react with tricalcium aluminate (C_3A) in the OPC, causing an expansion in the cement matrix. However, it can be said that the SCR-fly ash requires a resting time of approximately 20–30 days after collecting to attenuate ammonia concentration prior to being used as a cementitious additive in the cement and concrete industry.

$$Ca(OH)_2 + (NH_4)_2SO_4 \rightarrow CaSO_4 \bullet 2H_2O + 2NH_4OH \qquad (10.1)$$

SEM micrographs of 40% SCR-fly ash and 40% HC-fly ash mixtures are as presented in Fig. 10.6. Compact and dense structures were observed in both types of fly ashes with some unreacted fly ash particles. Nevertheless, it seemed that the

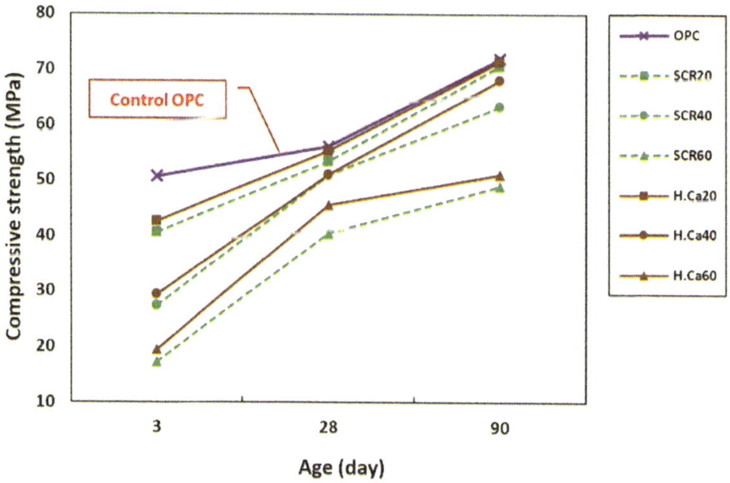

Fig. 10.5 Compressive strength of OPC-fly ash blended cement pastes with different FA types and different %FA additions

Fig. 10.6 Microstructures of OPC-blended fly ash of **a** 40% SCR-fly ash and **b** 40% HC-fly ash at their 28 days age

SCR40 showed higher porosity than the H.Ca 40 due to the possible expansion from the gypsum content.

10.5 Conclusion

1. Particle size and surface area of fly ash directly affected the physical properties (setting time and flowability) of the fresh cement paste. The smaller spherical particle of high calcium fly ash achieved a good hydration reaction, resulting in a shorter setting time and lower flow value than that of a larger particle of SCR fly ash.

2. The SCR-fly ash can be used as a cement additive like the typical high calcium fly ash with a slight reduction in compressive strength. It is highlighted that the SCR-fly ash must expose to the ambient atmosphere for at least 20 days to release the toxic ammonia. In this study, the SCR-fly ash was at 200 days-age from the collection date, and its ammonia content was decreased to the minimum level of 2.17 mg/kg.
3. The presence of too high ammonia content, e.g., over 50 mg/kg, could probably cause a formation of unwanted gypsum ($CaSO_4$), leading to an expansion of the cement and concrete in a later stage with scattered voids and internal microcracks. The overall strength of the cement and concrete may be significantly dropped.
4. The new type of fly ash from the selective catalytic reduction (SCR) process of the Mae Moh power plant in Lampang province can be used as an additive to reduce the amount of Portland cement consumption with good engineering properties. However, SCR-fly ash should be rested for a month to release the existed ammonia. This work could be an initial study for further utilizing of that SCR-fly ash.

Acknowledgments This work (Grant No. RGNS 63-078) was supported by the Office of the Permanent Secretary, Ministry of Higher Education, Science, Research and Innovation (OPS MHESI), Thailand Science Research and Innovation (TSRI). Also, this research work was partially supported by Chiang Mai University. The authors would like to express gratitude to the Department of Civil Engineering, Faculty of Engineering, Chiang Mai University (CMU), for providing equipment and facilities. Thanks to the Electricity Generating Authority of Thailand (EGAT) for the supply of fly ash. Also, the lovely and hardworking students, Jutarat Leetanarung, Supathida Maneewan, and Surunda Rueanwin, who kindly helped to conduct the experimental works.

References

ACI COMMITTEE 318 (1999) Building code requirements for structural concrete (ACI 318M-99) and commentary (ACI 318RM-99)

Alexander M, Bertron A, De Belie N (2013) Performance of cement-based materials in aggressive aqueous environments, vol 10. Springer, Berlin

Bualuang T, Jitsangiam P, Suwan T, Rattanasak U, Jakrawatana N, Kalapat N, Nikraz H (2021) Non-OPC binder based on a hybrid material concept for sustainable road base construction towards a low-carbon society. J Mater Res Technol 14:374–391

EGAT (2020) Electricity generating authority of Thailand. https://www.egat.co.th/index.php?option=com_content&view=article&id=2494&Itemid=117. Accessed 25 Feb 2022

Fisher BC, Blackstock T, Hauke D (1997) Fly ash beneficiation using an ammonia stripping process. In: Proceedings of the 12th international symposium on coal combustion-by products management and use, pp 65-1

Jitsangiam P, Suwan T, Pimraksa K, Sukontasukkul P, Chindaprasirt P (2021) Challenge of adopting relatively low strength and self-cured geopolymer for road construction application: a review and primary laboratory study. Int J Pavement Eng 22(11):1454–1468

Larrimore L (2002) Effects of ammonia from post-combustion NO_x control on ash handling and use. Fuel Chem Div Prepr 47(2):832–833

Lu C, Wang D, Zhu W, Wang F, Du Z, Zhu Y (2020) Performance evaluation and life management of SCR denitration catalyst. IOP Conf Ser: Earth Environ Sci (IOP Publishing) 495(1):012035

Maichin P, Suwan T, Jitsangiam P, Chindaprasirt P, Fan M (2020) Effect of self-treatment process on properties of natural fiber-reinforced geopolymer composites. Mater Manuf Process 35(10):1120–1128

Michalik A, Babińska J, Chyliński F, Piekarczuk A (2019) Ammonia in fly ashes from flue gas denitrification process and its impact on the properties of cement composites. Buildings 9(11):225

Suwan T, Jitsangiam P, Chindaprasirt P (2020) Influence of nano-silica dosage on properties of cement paste incorporating with high calcium fly ash Key Engineering. Materials (Trans Tech Publications Ltd.) 841:9–13

Wattanachai P, Suwan T (2017) Strength of geopolymer cement curing at ambient temperature by non-oven curing approaches: an overview. IOP Conf Ser: Mater Sci Eng 212(1):012014

Part III
Renewable Energy Technology and Energy Chemical Engineering

Chapter 11
Experimental Study on the Dynamic Responses of a 2 MW Cross-Shaped Multi-Column Spar Floating Offshore Wind Turbine

X. Ci, W. Li, Y. Lei, S. Gao, S. Zhang, and X. Y. Zheng

Abstract In this study, a new conceptual cross-shaped multi-column spar foundation is proposed for floating offshore wind turbines. This foundation consists of a reinforced concrete cross pontoon, a central steel column, and four side steel columns symmetrically distributed on pontoon ends. This foundation is meant for the economical wind power exploitation particularly in finite water depth ranging from 30 to 100 m. Using a 2 MW wind turbine in 100 m water as the prototype, comprehensive experimental studies are carried out at Tsinghua University Ocean Basin with a 1/20th scale under the Froude law to investigate the dynamic responses of this novel floater. In order to emulate the real mooring system, a truncated catenary mooring arrangement is optimized in the experiment. Model tests have been performed in different environmental loadings, including regular and irregular waves, steady winds, and their couplings to investigate the dynamic characteristics pertaining to structural safety. The obtained motion response amplitude operators (RAOs) and spectral responses demonstrate that this cross-shaped and multi-column spar wind turbine owns excellent dynamic performance in severe wave and wind conditions.

Keywords Floating offshore wind turbines · Conceptual design · Cross-shaped multi-column spar · Model tests · Dynamic responses

X. Ci · Y. Lei · S. Zhang · X. Y. Zheng (✉)
Institute for Ocean Engineering, Tsinghua Shenzhen International Graduate School, Shenzhen 518055, China
e-mail: zheng.xiangyuan@sz.tsinghua.edu.cn

X. Ci
Department of Mechanical Engineering, Tsinghua University, Beijing 100084, China

W. Li · S. Gao
Key Laboratory of Far-Shore Wind Power Technology of Zhejiang Province, Hangzhou 310014, China

Y. Lei
China Huaneng Clean Energy Research Institute, Beijing 102209, China

J. Baeyens et al. (eds.), *Proceedings of 2022 4th International Conference on Environment Sciences and Renewable Energy*, Environmental Science and Engineering, https://doi.org/10.1007/978-981-19-9440-1_11

11.1 Introduction

Offshore wind power is in the spotlight nowadays. The Global Offshore Wind Report 2021 (Joyce and Feng 2021) released by The Global Wind Energy Council (GWEC) announces that the global offshore wind power industry adds more than 6GW of installed capacity in 2020, and the total installed capacity has reached 35.3 GW, which accounted for 5% of total global wind capacity as of the end of 2020. At present, the construction of the offshore wind farm is mainly concentrated in the shallow area, using fixed foundations. Over recent 5 years, several concepts of floating offshore wind turbine (FOWT) have been successfully developed to adapt to water depth deeper than 50 m.

There are three main types of the FOWT: spar-buoy, tension leg and semisubmersible. The water depth of spar-buoy foundation is usually over 200 m. Bjørn et al. (Skaare et al. 2015) carried out a comparative dynamic analysis between full-scale tests and numerical simulations from the floating wind turbine Hywind Demo, showing good agreements. Christian et al. (Cermelli et al. 2018) compared the experimental results of WindFloat 1 platform with the numerical model developed by Principle Power. Beyer et al. (2015) and Borisade et al. (2016) investigated the coupled MBS-CFD simulation of the IDEOL floating offshore wind turbine foundation. The model tests are used to validate the reliability of the coupled MBS-CFD method. Karimirad et al. (Karimirad and Michailides 2015) utilized the aero-hydro-servo-elastic numerical modelling to investigate the dynamics responses of a V-shaped semisubmersible. Their findings showed that such a semisubmersible concept is well suited for complex environmental conditions. Considering that a number of semisubmersible concepts have been proposed and applied in commercial floating offshore wind farms with 50–300 m water depth, Liu et al. (2016) comprehensively compared these concepts from three aspects, i.e. of design, experimental study, and numerical study. Technically, the stability of a semisubmersible wind turbine is inferior to a spar-buoy wind turbine. Though the Hywind spar wind turbines have been successfully deployed in Scotland water of depth greater than 100 m for over 4 years, in shallow and finite water depths, it is still noteworthy to revisit such a design for the sake of its excellent stability and seakeeping performance. Nonetheless, the original Hywind design has to undergo a great change whereas a strategy is taken the respective advantages of spars and semisubmersibles.

The conceptual design of foundation is mainly concerned with increasing structural reliability and reducing costs, which has been highlighted in a recent study by Moan et al. (2020). Nowadays almost all FOWTs use steel materials to ensure the strength of the structure, while some concrete foundation FOWTs like IDEOL are also developed in the commercial project. These concrete foundations have demonstrated great prospects with outstanding durability and cost-effectiveness.

In view of the current trends in offshore wind industry, this paper proposes a new conceptual cross-shaped multi-column spar FOWT with both steel and concrete materials, particularly for the wind power exploitation in coastal regions of water depth ranging from 30 to100 m. The design parameters are given, and the results of model tests are shown.

11.2 Conceptual Design

The concept of cross-shaped and multi-column spar FOWT is shown in Fig. 11.1. A 2 MW wind turbine is installed on the top of the central column. The mooring lines are connected to the four cross ends of the foundation.

11.2.1 A 2 MW Wind Turbine and Tower

The wind turbine of the prototype is W2000-99-80 2 MW, which is designed by Shanghai Electric Windpower Equipment CO., LTD. In the model tests, a commercial wind turbine (Fig. 11.2) is used to emulate the wind loads on the prototype,

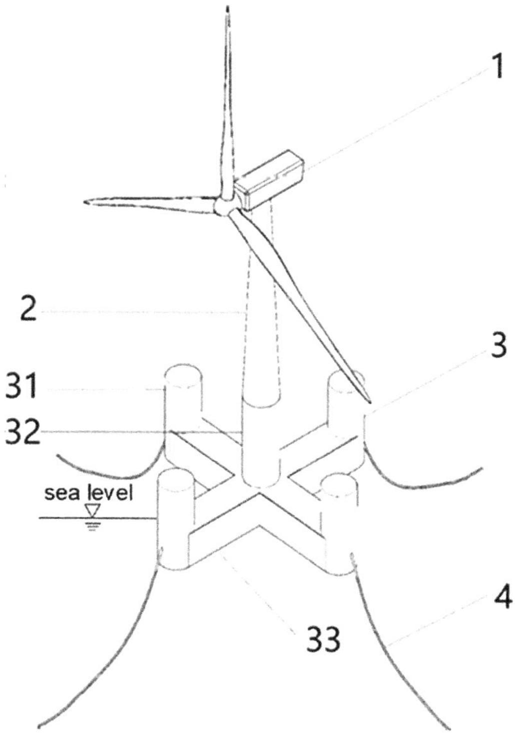

Fig. 11.1 Cross-shaped and multi-column spar FOWT. (1-wind turbine, 2-tower, 3-foundation, 31-side columns, 32-central column, 33-cross pontoon, 4-mooring lines)

Fig. 11.2 A commercial wind turbine employed in experiments

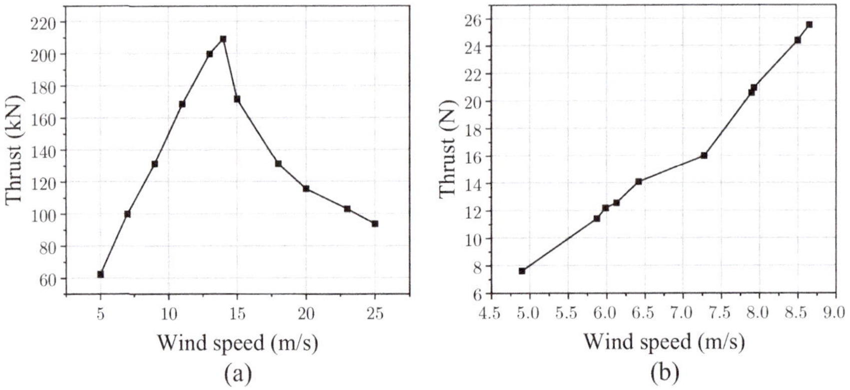

Fig. 11.3 Thrust curves of prototype (**a**) and model (**b**)

considering the thrust similitude principle with 1/20th scale under the Froude law. The thrust curves of prototype and model are shown in Fig. 11.3.

The model tower is made of aluminium alloy. Its outer diameter, thickness and the length are respectively 58 mm, 10 mm, and 3762 mm. Two acceleration sensors and a six-axis force transducer are mounted on the upper part of the model tower. The properties of the model turbine and model tower are given in Table 11.1. In this table, the centre of gravity and hub height (80 m) are measured from the tower bottom. It can be seen that the mass properties and the natural frequency of the model agree well with the target values.

11.2.2 Cross-Shaped and Multi-Column Spar Foundation

The spar foundation consists of a reinforced concrete cross pontoon, a central steel column, and four side steel columns symmetrically distributed on pontoon ends. The geometric parameters of the foundation are listed in Table 11.2. The cross

Table 11.1 Table captions should be placed above the tables

	Prototype	Model target	Actual Model
Hub height (m)	80	4	3.98
Overall mass (kg)	297,754	36.31	35.59
Overall CG (m)	49.54	2.48	2.54
1st vibration natural frequency (Hz)	0.438	1.959	1.958

Table 11.2 The geometric parameters of the foundation

Structural members	Sectional diameter or breadth × height (m)	Length (m)	Thickness (m)
Central column	9.50	18.63	0.03
Side column	8.50	17.63	0.02
Pontoon	11.50 × 6.38	50.25	0.30

shape ensures convenient access of support vessels to the wind turbine hub during marine operation and maintenance. Also, such a foundation facilitates construction and increases moment of inertia of the floater.

The model's foundation (Fig. 11.4) is fabricated with aluminum alloy, which satisfies the geometric similarity. Twenty-four steel counterweights are placed symmetrically in the cavity of the pontoon as the ballast to adjust the moment of inertia and model's CG. The properties of the prototype and the model are summarized in Tables 11.3 and 11.4. The CG and centre of buoyancy (CB) of the model are calculated from the still water level. The moment of inertias for roll, pitch and yaw are measured about the CG. As CG is 1.33 m lower than CB, this design therefore takes the advantage of spar for being unconditionally stable.

11.2.3 *Truncated Mooring System*

The experimental studies are carried out at The Ocean Basin of Tsinghua University Shenzhen International Graduate School. As the water depth for the prototype is 100 m, the 1/20th scale model was therefore designed for the water depth of 5 m in this basin. Due to the size limitation of the basin, a truncated mooring system has to be developed according to the Molins (Molins et al. 2015). The layout of the whole system is in Fig. 11.5.

The truncated mooring system consists of eight catenary lines. As illustrated in Table 11.5, Lines 1–4 are made up of two different stainless chains while lines 5–8 use a uniform stainless chain. To measure the tension forces on catenary lines during the experiment, two tension sensors are placed between line 1, 7 and the model foundation.

Fig. 11.4 The model in the ocean basin

Table 11.3 The properties of the prototype and the model

	Prototype	Model target	Model
Draft (m)	19	0.95	0.95
Freeboard (m)	6	0.3	0.3
Mass (tons)	9333.15	1.14	1.16
CG (m)	−13.96	−0.7	−0.72
CB (m)	−12.63	−0.63	−0.65
Iroll about CG (kg·m^2)	3.32E9	1012.87	1005.91
Ipitch about CG (kg·m^2)	3.32E9	1012.87	1005.91
Iyaw about CG (kg·m^2)	2.76E9	841.56	818.42
Displacement (m^3)	9105.51	1.14	1.16

Table 11.4 Natural periods of seakeeping motions (prototype)

Freedom degrees	Surge	Heave	Pitch
Natural period(s)	95.45	19.44	17.68

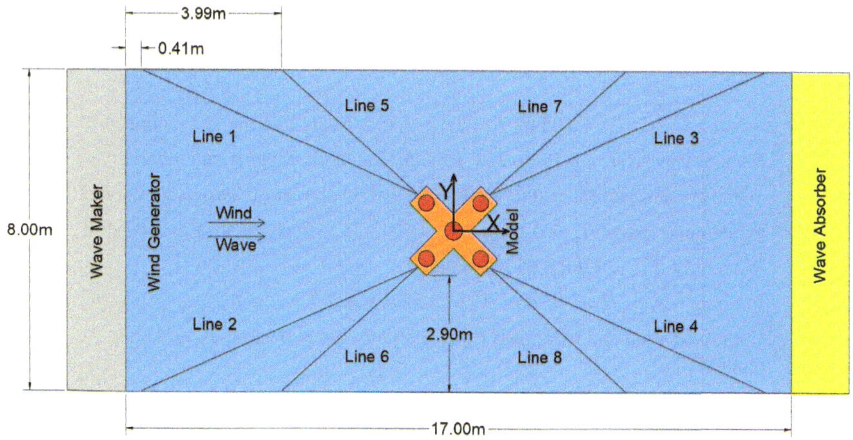

Fig. 11.5 Layout of the moored physical model in ocean basin

Table 11.5 Properties of the mooring lines in model tests

	Lines 1–4	Lines 5–8
Length (m)	Upper segment: 6.00	7.32
	Lower segment: 3.06	
Mass density in water (kg/m)	Upper segment: 0.21	0.42
	Lower segment: 1.01	

11.3 Model Tests

After a series of wave and wind conditions are calibrated, the model is placed in a predetermined position. Key physical quantities to be measured in the model tests include the response amplitude operators (RAOs) of 6-DOFs seakeeping motions, tension forces of line 1 and line 7, accelerations at the top of the tower, forces at the connection between the turbine and tower. The selected load cases are listed in Table 11.6 that includes waves (regular and irregular), wind, and their coupling. In this study, the directions of wind and waves are positive along X-axis.

The regular waves and white noise waves are generated to obtain the motion RAOs of the model. Two series of regular waves of different wave heights (LC1 & LC2) and two white noise waves with different significant wave heights (LC3 & LC4) are adopted to determine the displacement RAOs in freely floating and mooring states. Wind only tests (LC5) are carried out to determine the motion of the model when turbine is operating. LC6 represents the 50-year extreme sea state. Lastly, the combined wind and wave test (LC7) is implemented, of which the irregular waves are represented by a JONSWAP wave spectrum. The spectral parameters H_s, T_p and γ are given in Table 11.6. The duration LC7 is 3 h in the prototype after removing the transient response.

Table 11.6 Load cases in model tests

Case no	Description
LC1	Regular waves: $H = 0.05$ m, $T = 1.12$–3.35 s, freely floating and moored
LC2	Regular waves: $H = 0.10$ m, $T = 1.57$–3.35 s, freely floating and moored
LC3	White noise wave: $H_s = 0.05$ m, $T = 1.12$–3.35 s, freely floating and moored
LC4	White noise wave: $H_s = 0.10$ m, $T = 1.57$–3.35 s, freely floating and moored
LC5	Wind only: $V_w = 4.90$–8.65 m/s, moored
LC6	Irregular waves: $H_s = 0.26$ m, $T_p = 2.58$ s, $\gamma = 2.2$, moored
LC7	Combined wind and wave test: $V_w = 8.65$ m/s, $H_s = 0.21$ m, $T_p = 2.18$ s, $\gamma = 2.1$, moored

11.4 Results and Discussion

11.4.1 Tests for Motion RAOs

The surge RAOs of the moored spar FOWT are shown in Fig. 11.6a. For common regular waves with periods between 5 and 15 s, the surge RAOs are less than 1.25 m/m. The curves of H = 1 m and H = 2 m match well in short period waves (5 and 13 s). However, their difference gradually becomes larger when T is greater than 13 s. To reflect the characteristics of motions more comprehensively, the mean surge RAOs which is defined by the ratio of the mean displacement of surge to the wave height are given in Fig. 11.6b. The mean surge RAOs of H = 2 m are significantly larger than that of H = 1 m, especially in short wave periods, in that nonlinear wave loads on the floater like the Morison drag force are roughly proportional to the square of the wave height.

The heave RAOs and the pitch RAOs are given in Fig. 11.6c and d under moored and freely floating states. In moored state, the heave RAOs are less than 0.5 m/m and the pitch RAOs are less than 1.3 deg/m within the common wave periods 5–15 s, showing very good seakeeping performances. The heave RAOs of moored state are slightly smaller than the freely floating state, indicating the restriction imposed by anchor chains. There is also a good consistency between the pitch RAOs of moored and freely floating states for 5–11 s. In Fig. 11.6, for T > 13 s, RAOs for H = 2 m are distinct from those for H = 1 m. These differences are caused by the nonlinearity of the mooring system and the viscous effect of Morison drag force (Lei et al. 2021).

11.4.2 Wind Only Tests

Wind only tests focus on surge and pitch motions of the spar FOWT, as well as the shear force F_x on the top of the tower. The results are shown in in Fig. 11.7. The wind

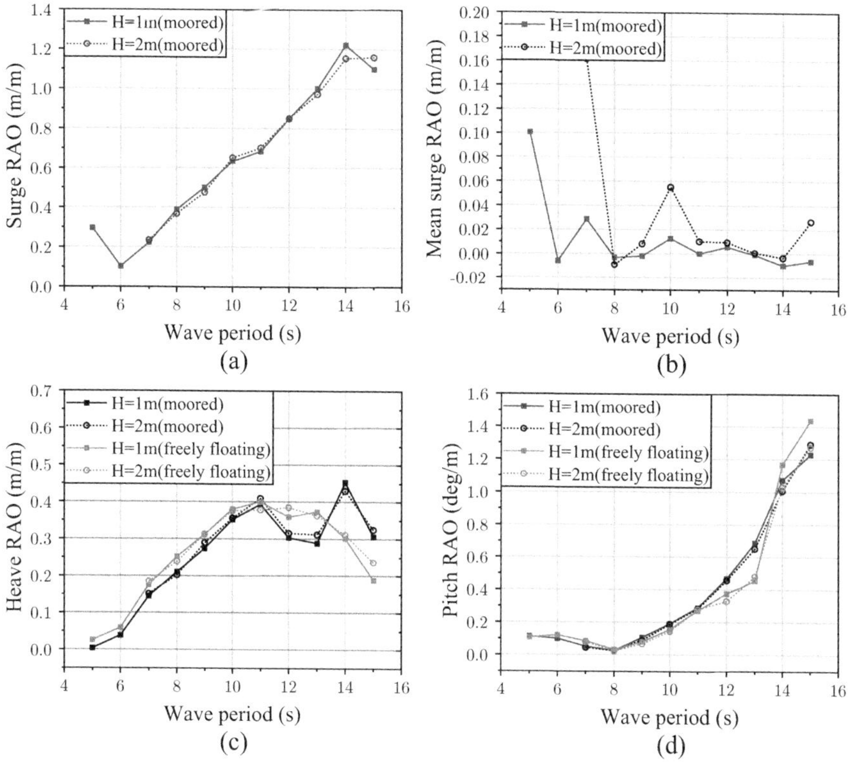

Fig. 11.6 RAOs of the spar FOWT

speeds are selected from the thrust curve of the W2000-99-80 2 MW wind turbine (Fig. 11.3). When the wind speed is 14 m/s, mean F_x is 145.98 kN, mean surge is 2.25 m and mean pitch is 1.42°.

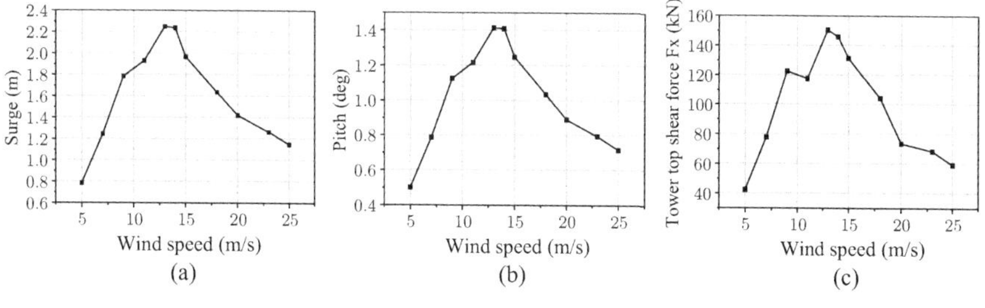

Fig. 11.7 The mean values of surge (**a**), pitch (**b**), and F_x (**c**) in prototype

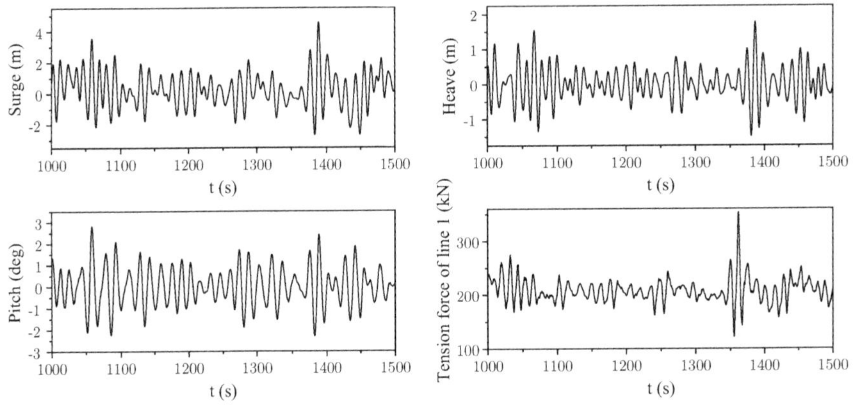

Fig. 11.8 Partial response histories of surge, heave, pitch, and TLine1 for the 50-year sea state (LC6)

Table 11.7 The response statistics of prototype in the 50-year sea state (LC6)

	Surge (m)	Heave (m)	Pitch (m)	T_{Line1} (kN)
Mean	0.24	0.01	0.02	207.83
Std	1.03	0.46	0.83	19.62
Max	4.56	1.79	3.11	353.58
Min	−2.90	1.57	−2.47	121.53

11.4.3 Irregular Wave Tests

Under the 50-year extreme wave state (LC6), the partial response time series of the surge, heave, pitch, and tension force of Line 1 are given in Fig. 11.8. Their statistics are shown in Table 11.7. For the prototype, the maximum surge is 4.56 m, the maximum heave is 1.79 m, the maximum pitch is 3.1°, and the maximum tension force of Line 1 is 353.58 kN with pretension of 201.05 kN.

The power spectra of surge, heave, pitch, and T_{Line1} are given in Fig. 11.9. All spectral peaks at the wave frequencies can be clearly observed. Also, low frequency resonant responses are also pronouncedly displayed in heave and surge spectra. By contrast, the pitch motion whose natural period is 17.68 s is subjected to long wave periods.

11.4.4 Combined Wind and Wave Test

Figure 11.10 illustrates the dynamic responses of surge, heave, pitch, tension force of Line 1, and F_x under the combined wind and wave extreme condition (LC7).

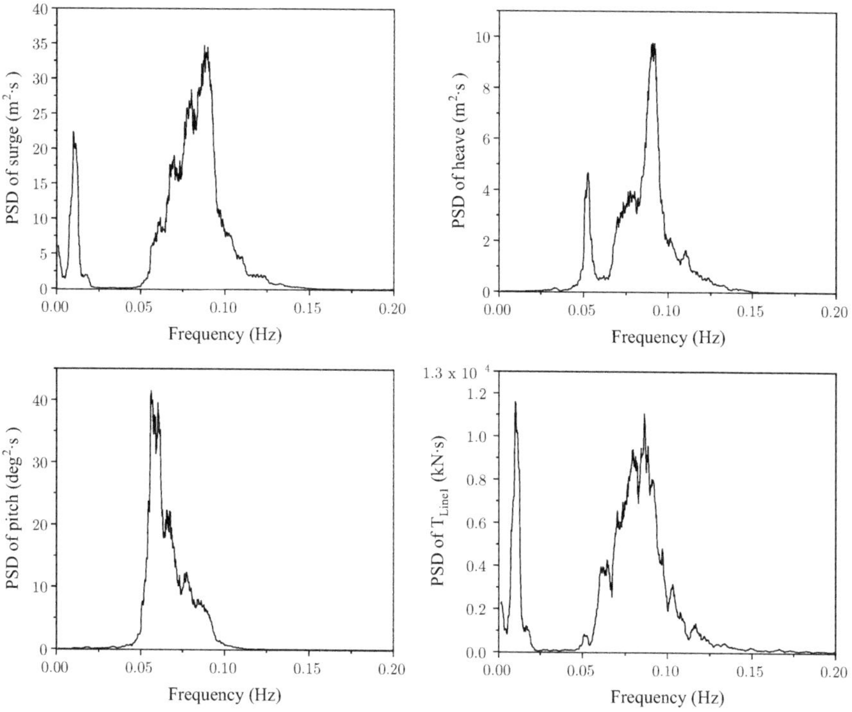

Fig. 11.9 The response power spectra of prototype in the 50-year sea state (LC6)

Obviously, F_x contains high frequency components. The corresponding response statistics for this sea state are presented in Table 11.8. The maximum surge is 5.16 m, the maximum heave is 1.33 m, the maximum pitch is 2.65°, the maximum T_{Line1} is 317.95kN and the maximum F_x is 362.08kN.

The power spectra of surge, heave, pitch, T_{Line1}, and F_x are shown given in Fig. 11.11. The responses at wave frequencies and the respective low natural frequencies are well manifested. However, due to fast blade speed, the vibration of the tower at its natural frequency is also excited during this test, i.e., a significant spectral peak at 0.44 Hz in the spectrum of F_x.

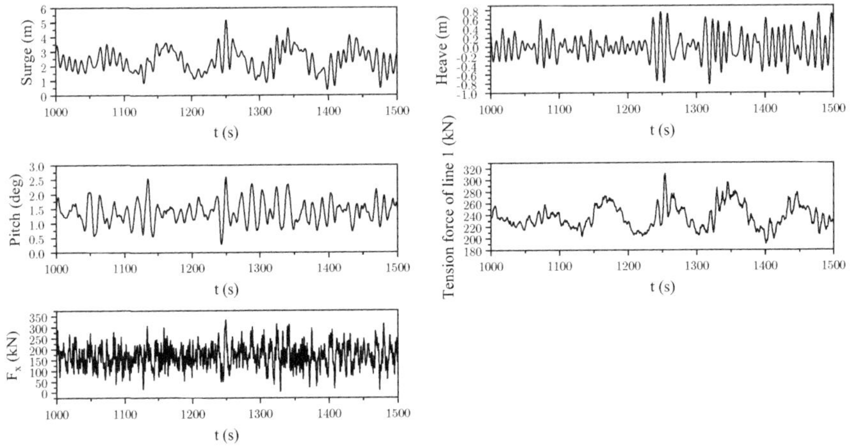

Fig. 11.10 The partial response time series in the combined wind and wave condition (LC7)

Table 11.8 The response statistics in the combined wind and wave condition (LC7)

	Surge (m)	Heave (m)	Pitch (m)	T_{Line1} (kN)	F_x (kN)
Mean	2.40	0.01	1.37	238.76	165.48
Std	0.68	0.31	0.41	15.86	55.93
Max	5.16	1.33	2.65	317.95	362.08
Min	0.38	−0.96	0.01	190.10	−43.29

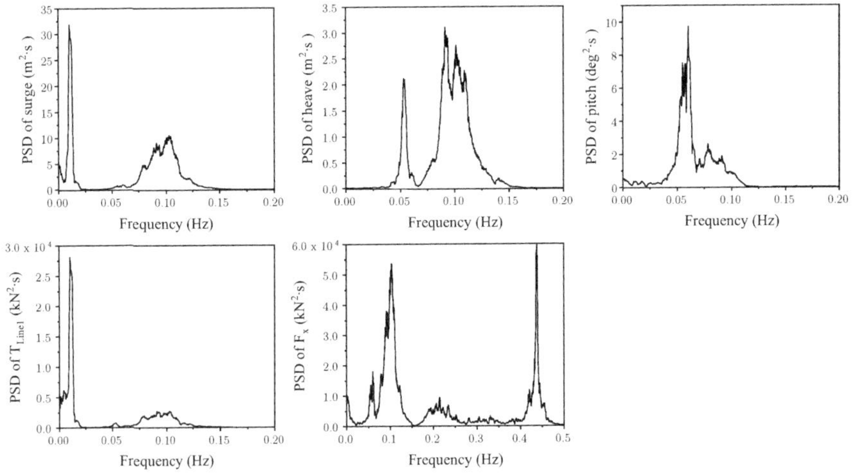

Fig. 11.11 The response power spectra responses in the combined wind and wave condition (LC7)

11.5 Conclusion

This paper proposes a new conceptual cross-shaped multi-column spar FOWT for 30–100 m water depth. Its seakeeping and dynamic responses are investigated by a series of comprehensive model tests with the 1/20th scale. The experimental results demonstrate very good seakeeping and dynamic performances of this floater under various sea states. Thereinto, the heave RAOs are less than 0.5 m/m and the pitch RAOs are less than 1.3 deg/m within the common range of regular wave periods 5–15 s in the moored state. The response time histories and their spectra under extreme wind and wave conditions also verify the excellent dynamic performance of such a new floating wind turbine, i.e., the maximum heave and pitch are respectively 1.33 m and 2.65°. Though this conceptual design is meant for a 2 MW wind turbine, it is still applicable to a large megawatt wind turbine in water depth exceeding 100 m.

Acknowledgements The financial supports received from 2020 Open Fund Projects of Key Laboratory of Far-shore Wind Power Technology of Zhejiang Province and China National Science Foundation Program (52071186) are greatly acknowledged.

References

Beyer F, Choisnet T, Kretschmer M, Cheng PW (2015) Coupled MBS-CFD simulation of the IDEOL floating offshore wind turbine foundation compared to wave tank model test data. In: The Twenty-fifth (2015) international ocean and polar engineering conference, Hawaii

Borisade F, Choisnet T, Cheng PW (2016) Design study and full scale MBS CFD simulation of the IDEOL floating offshore wind turbine foundation. J Phys Conf Ser 753

Cermelli C, Leroux C, Díaz Domínguez S, Peiffer A (2018) Experimental measurements of WindFloat 1 prototype responses and comparison with numerical model. American Society of Mechanical Engineers, Spain

Joyce L, Feng Z (2021) Global offshore wind report 2021: global wind energy council

Karimirad M, Michailides C (2015) V-shaped semisubmersible offshore wind turbine: an alternative concept for offshore wind technology. Renew Energy 83:126–143

Lei Y, Zheng XY, Li W, Zheng HD, Zhang Q, Zhao SX et al (2021) Experimental study of the state-of-the-art offshore system integrating a floating offshore wind turbine with a steel fish farming cage. Mar Struct 80

Liu Y, Li S, Yi Q, Chen D (2016) Developments in semi-submersible floating foundations supporting wind turbines: a comprehensive review. Renew Sustain Energy Rev 60:433–449

Moan T, Gao Z, Bachynski EE, Nejad AR (2020) Recent advances in integrated response analysis of floating wind turbines in a reliability perspective. J Offshore Mech Arctic Eng 142(5)

Molins C, Trubat P, Gironella X, Campos A (2015) Design optimization for a truncated catenary mooring system for scale model test. J Mar Sci Eng 3(4):1362–1381

Skaare B, Nielsen FG, Hanson TD, Yttervik R, Havmøller O, Rekdal A (2015) Analysis of measurements and simulations from the Hywind Demo floating wind turbine. Wind Energy 18(6):1105–1122

Chapter 12
New Promising Modified Activated Carbons for CH_4 and CO_2 Adsorption

G. Iragena Dushime, J. Bachelart, K. Abou Alfa, C. Matei Ghimbeu, C. Hort, and V. Platel

Abstract This work focuses on methane (CH_4) and carbon dioxide (CO_2) pure gases adsorption using commercial activated carbon (CNR-115), and its modified activated carbons: the one obtained via oxidation (CNR-115-ox) and the other given by oxidation followed by ammonium impregnation (CNR-115-ox-am). A homemade setup was used for isotherms determination at 303.15 K and 0–3 MPa. At 3 MPa, CO_2 and CH_4 uptakes were: CNR-115 (12.05 and 5.18 mmol/g respectively) > CNR-115-ox (7.79 and 3.16 mmol/g respectively) > CNR-115ox-am (4.05 and 1.53 mmol/g respectively). Hence: (a) CNR-115 with high BET surface area (1714 m^2/g) adsorbs higher amount of gases than CNR-115-ox (929 m^2/g), which also adsorbs more gases than CNR-115-ox-am with the least BET surface area (352 m^2/g), and (b) CO_2 with smaller molecular size (330 pm) is more adsorbed on activated carbons surface than CH_4 (380 pm). Further, the Langmuir model was used for adsorption description and the results are well comparable to the ones reported in the literature. Lastly, an interesting side of CNR-115-ox-am was discussed, and this activated carbon was found to be promising for CH_4 enrichment from CH_4/CO_2 mixtures due to its low CH_4 uptake.

Keywords Activated carbons · Adsorption · Carbon dioxide · Methane

G. I. Dushime (✉) · J. Bachelart · K. A. Alfa · C. Hort · V. Platel
Université de Pau et des Pays de l'Adour/E2S UPPA, Laboratoire de Thermique, Energetique Et Procedes, EA1932, 64000 Pau, France
e-mail: grace.iragena-dushime@univ-pau.fr

C. M. Ghimbeu
Institut de Sciences des Matériaux de Mulhouse, CNRS UMR 7361 UHA, 15 Rue Jean Starcky, 68057 Mulhouse, France

J. Baeyens et al. (eds.), *Proceedings of 2022 4th International Conference on Environment Sciences and Renewable Energy*, Environmental Science and Engineering, https://doi.org/10.1007/978-981-19-9440-1_12

12.1 Introduction

The global world climate change and energy sector development are two main keys factors to promote the use of renewable energies. Historically, the use of gas for energy production began at the end of the nineteenth century. The sector was then improved and underwent many transformations (Gasquet 2020). Indeed, the first-generation gas called the city gas was obtained from coal pyrolysis and was then replaced by natural gas due to the fact that, the latter is naturally formed in some porous rocks and extracted by drilling instead of pyrolysis which also requires the use of energy. Currently, biogas (i.e. the third generation gas) is gaining too much attention over coal and natural gas because the former is a renewable energy contributing in climate protection by reducing the emissions of greenhouse gases such as carbon dioxide (CO_2) (Vondra et al. 2019). Biogas (i.e. The mixture of CH_4 (30–75%) and CO_2 (15–50%)) is produced from a process called anaerobic digestion through biochemical degradation of organic wastes materials (Boulinguiez and Cloirec 2015). However, the presence of carbon dioxide is not crucial for the use of biogas, as it lowers its heating value. This is the main reason of upgrading the biogas in order to produce biomethane which can therefore be used: (i) for energy production, ii) as fuel for vehicles, and iii) natural gas pipelines injection. Concerning pipelines injection of biomethane, the gas must fit pipelines-quality requirements (i.e. $CO_2 < 3\%$, and $CH_4 > 97\%$) (Olivier 2015). Biomethane is almost similar to natural gas in terms of calorific heat (>8.55 and 11.4 kWh/Nm^3 for biomethane and natural gas respectively (Rogulska et al. 2018)), the difference between them is based on CO_2 emissions. For example, since 2017 the direct emissions of carbon from biomethane in France (23 gCO_2/kWh) was about ten times less than the one from natural gas (244 gCO_2/kWh) (Gasquet 2020), this explains the reason to prefer biomethane than natural gas. However, the price of biomethane in France (~100 €/MWh) is still higher compared to the price of natural gas (<20 €/MWh) (Tilagone and Lecointe 2015), this is why many researches are increasingly conducted dealing with ways to reduce this price. Reducing the costs of biogas upgrading is one of ways to reduce the biomethane price. This work highlights the role of activated carbons (ACs) in biogas upgrading. Different technologies are known for biogas upgrading: Pressure Suing Adsorption (PSA), water scrubbing, amine scrubbing, membrane, cryogenic, and biological treatments. However, some of those technologies are pricey, others are energivores, and others emit chemicals enhancing climate-damaging. PSA technology is preferred due to its low costs and low energy requirements (Kapoor et al. 2019). Moreover, PSA is based on the use of solid adsorbent materials such as activated carbons, zeolites, and metal organic frameworks to remove CO_2 via adsorption process; this makes it applicable over a wide range of temperatures and pressures (Speight 2019). Among adsorbents for CO_2 removal, activated carbons are considered as the best, thanks to their good textural properties, their good stability under acidic and basic conditions (El-Shafey et al. 2016), their low costs, and their ease of regeneration (Pu et al. 2021). ACs dispose high specific surface areas (400–3500 m^2/g) and large pores sizes (1.0–4.0 nm) (Sun et al. 2016), leading to high CO_2 adsorption capacity but low selectivity (Ribeiro et al. 2008). However, the selectivity of activated carbons can be optimized through surface modification (i.e. introducing chemical or/ and physical materials) (Zulkurnai et al. 2017). This work provides insight regarding the influence of activated carbons modification on CH_4 and CO_2 adsorption. For that concern, a homemade manometric setup was used, and a sufficient number of experiments were conducted on the adsorption of pure gases (CH_4 and CO_2) by commercial CNR-115 and two new modified activated carbons (CNR-115-ox and CNR-115-ox-am) at 303.15 K and the pressure range of 0–3 MPa. To describe the behavior of CH_4 and CO_2 on a set of activated carbons used, the Langmuir two parameters model was used and the results are well comparable to the ones are well comparable to the ones in the literature.

12.2 Materials and Methods

12.2.1 Commercial and New Modified Activated Carbons (Family of CNR-115)

Three activated carbons used in this work: commercial (CNR-115) and two new modified (CNR-115-ox and CNR-115-ox-am) were purchased at Mulhouse Materials Science Institute (IS2M, France). The surface chemistry of commercial CNR-115 was modified through two successive steps: (a) oxidation under air at 400 °C for 2 h to increase the number of oxygenated groups. Activated carbon produced in this step is called CNR-115-ox, and (b) ammonium vapor (NH_4OH) impregnation at 298 K for 36 h to increase the amount of nitrogenated groups. The resulted activated carbon in this step is called CNR-115-ox-am.

12.2.2 Activated Carbons Characterization

Textural properties of activated carbons were characterized by means of nitrogen adsorption/desorption measurements using a micrometrics ASAP2020 under 77 K. BET (Brunauer, Emmet and Teller) model was used for specific surface area (SSA) determination. The total pore volume (V_t) was obtained for P/P_0 equals to 0.99. Dubinin-Radushkevich model was used to calculate the micropore volume (V_{micro}), whereas the mesopore volume (V_{meso}) was given by the difference between V_t and V_{micro}. SAIEUS software (Micromeritics) and 2D-NLDFT (Non Local Density Functional Theory (Jagiello and Olivier 2013)) were applied to determine the pore size distribution (PSD) which served in the average pore width (L_0) determination through the Eq. (12.1).

Table 12.1 Textural properties of CNR-115, CNR-115-ox and CNR115-ox-am (Peredo-Mancilla et al. 2018)

Activated carbon	S_{BET} (m^2g^{-1})	V_{micro} (cm^3g^{-1})	V_{meso} (cm^3g^{-1})	V_{tot} (cm^3g^{-1})	L_0 (nm)
CNR-115	1714	0.64	0.31	0.95	1.10
CNR-115-ox	929	0.36	0.09	0.45	0.96
CNR-115-ox-am	352	0.14	0.07	0.21	0.78

$$L_0 = \frac{1}{N}\sum_{i=1}^{N} \frac{w \cdot dV/dW}{dV/dW} \quad (12.1)$$

where L_0 is the average pore diameter (nm), w is the pore width (nm), dV/dW is the differential pore volume per pore width (cm^3 nm^{-1}), and N is the number of data points. The Table 12.1 presents the textural properties obtained for activated carbons of the CNR-115 family: CNR-115, CNR-115-ox and CNR115-ox-am. In the Table 12.1, it can be observed that the surface modification process remarkably reduces the BET surface of activated carbons of the CNR-115 family in the following sequence: CNR-115 (1714 m^2 g^{-1}) > CNR-115-ox (929 m^2 g^{-1}) > CNR-115-ox-am (352 m^2 g^{-1}). The total pore volume has also decreased (more than half) after each modification process: CNR-115 (0.95 cm^2 g^{-1}) > CNR-115-ox (0.45 cm^2 g^{-1}) > CNR-115-ox-am (0.21 cm^2 g^{-1}). Micro and meso-pore volumes, and average pore size were also impacted. However, the decrease in BET surface area is considered as normal result of modification process due to the fact that, the latter includes the temperature rise and thus the mass vaporization. The decrease in porosity can be associated with pore blockage that takes place during modification where the loading of oxygen and nitrogen functional groups may block some pores (Madzaki et al. 2016).

On the other hand, the N_2 adsorption/desorption and 2D-NLDFT pore size distribution (Figs. 12.1 and 12.2 respectively) behave as type I isotherm because: (i) the adsorbed volume increases at low pressures, and (ii) there is the plateau formation at higher pressures.

12.2.3 Experimental System

Adsorption isotherms were determined by using manometric setup presented on the Fig. 12.3, at the temperature of 303.15 K and the pressure range of 0–3 MPa.

The system is composed of two basis elements which are the dosing and adsorption cells. To measure the pressure inside the cells, MKS pressure transducer baratron type 121 A (with 0.01% uncertainty from vacuum to 3.3 MPa) was used. The dosing and adsorption cells are separated by valves, whose also allow minimizing the dead space volume. Isothermal condition of the system was maintained at 303.15 K by wrapping the cells with a heating wire controlled by a Eurotherm 3208 PID regulator,

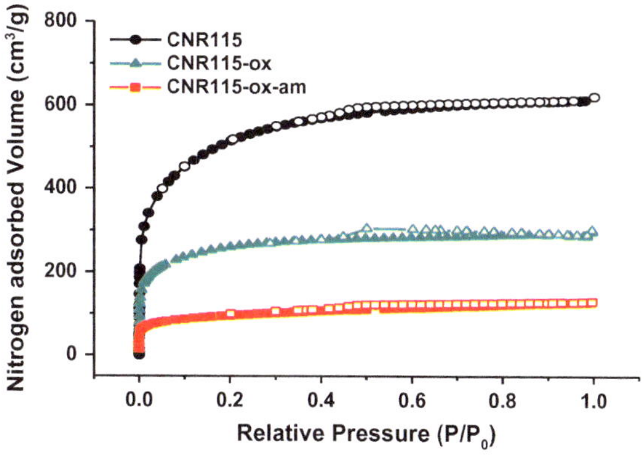

Fig. 12.1 N_2 adsorption/desorption of CNR-115, CNR-115-ox and CNR-115-ox-am (Peredo-Mancilla et al. 2018)

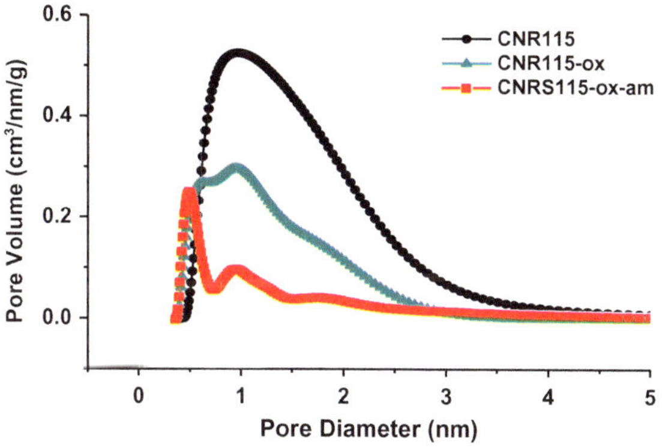

Fig. 12.2 Pore size distribution of CNR-115, CNR-115-ox and CNR-115 ox-am (Peredo-Mancilla et al. 2018)

and was verified by using two thermocouples fixed on the cells. The system is also connected to the vacuum pump for regeneration process.

12.2.4 Volume Calibration

The dosing and adsorption cell volumes of the empty system were previously calculated by Deneb et al. (Peredo Mancilla 2019) by means of gravimetric calibration method, and the values obtained were 30.8 and 19.05 cm^3 for dosing and adsorption volumes respectively. Determination of the volume occupied by adsorbent is then the preliminary step which must be done once a new mass of adsorbent sample is

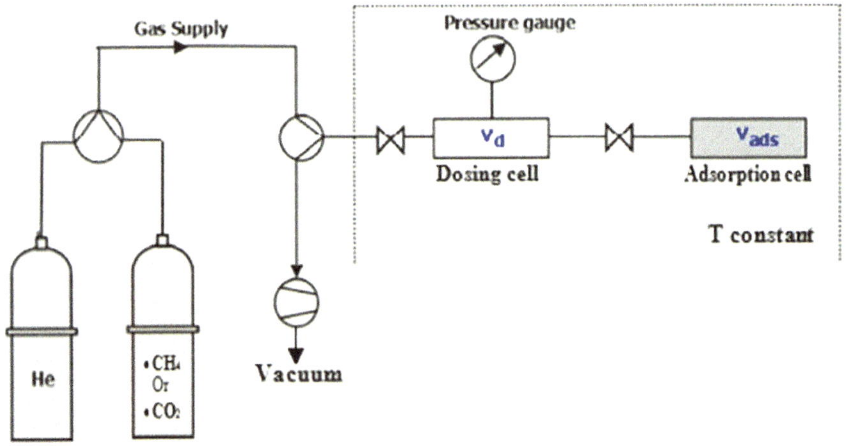

Fig. 12.3 High-pressure manometric device for pure gases adsorption (adapted from Peredo Mancilla 2019)

introduced into the system. This volume is slightly different from the one of the empty cell due to the fact that, the small volume is occupied by adsorbent. To do so, a small amount of activated carbon (~1 g) was placed inside the adsorption cell (after vacuum at 90 °C for 8 h). When the isothermal conditions at the experimental temperature (303.15 K) was reached, the small pressure of helium gas (i.e., inert gas which is not adsorbed by activated carbons) was introduced into the dosing cell. After the pressure stabilization, the initial pressure (P_i^i) was noted and the gas was expanded into the whole system (Bessières et al. 2005). At that stage, when the new stabilization of pressure was obtained, the final pressure (P_f^i) was taken and the first point (i = 1) of the dead space volume (V_m) of adsorption cell was calculated by means of the ideal gas equation of state using the Eq. (12.2). Other doses of helium were then introduced successively in the dosing cell and expanded into the adsorption cell. The dead-space volume for the ith step was therefore expressed by using the Eq. (12.3), and the final dead space volume was given by the average of all the measurements.

$$V_m = V_d((P_i/P_f) - 1) \tag{12.2}$$

$$V_m^i = V_d\left(\left(P_i^i - P_f^i\right)/\left(P_f^i - P_f^{i-1}\right)\right) \tag{12.3}$$

12.2.5 CH_4 and CO_2 Pure Gases Adsorption Measurements

To determine CH_4 and CO_2 pure gases adsorption isotherms, the main steps of experimental methodology are the following: (i) a small activated carbon sample was measured and introduced directly into the adsorption cell. In this work, the amount of around 1 g was chosen in order to get enough available adsorption area (50 m^2 is the minimum surface required) (Mouahid et al. 2011). (ii) The experimental system was regenerated under vacuum at 90 °C for 8 h in order to remove all the gases fixed at the surface of activated carbon. (iii) The third step is to calculate the volume occupied by the adsorbent (as detailed above). (iv) The experimental temperature (303.15 K) was then settled. (v) Lastly, the successive doses of gas (CH_4 or CO_2) were introduced and expanded into the dosing and adsorption cells respectively, and the initial and final pressures were noted after reaching the equilibrium. The number of adsorbed moles for the first point of the adsorption isotherm (n_{ads}^1) was calculated by using the Eq. 12.4, and the others points of the adsorption isotherm (n_{ads}^i) were given by the Eq. (12.5).

$$n_{ads}^1(T, P_1) = (V_d \rho_i) - ((V_d + V_m)\rho_f) \tag{12.4}$$

$$n_{ads}^i(T, P_i) = \left(V_m\, \rho_f^{i-1} + V_d \rho_i^i\right) - \left((V_d + V_m) \cdot \rho_f^i\right) \tag{12.5}$$

where ρ_i and ρ_f (g cm^{-3}) are the molar density at initial and final pressures respectively. The values of ρ_i and ρ_f at the experimental temperature and pressures conditions were taken from the NIST (National Institute of Standards and Technology) database.

12.2.6 Langmuir Fitting Model

To study the adsorption behavior of pure CH_4 and CO_2 on the mentioned adsorbents, Langmuir two-parameter model (which takes into account the adsorbed phase) was used as in Eq. (12.6) (Ortiz Cancino et al. 2017).

$$n_{ads}^{excess} = n_L \frac{P}{P + P_L}\left(1 - \frac{\rho_{g\,(P,T)}}{\rho_{ads}}\right) \tag{12.6}$$

where n_{ads}^{excess} (mol kg^{-1}) is the amount of gas adsorbed at the working pressure P (MPa), P_L (MPa) is the Langmuir pressure, ρ_g (kg m^{-3}) is the gas density, and ρ_{ads} (kg m^{-3}) is the adsorbed phase density. In this work, the values of ρ_{ads} are fixed to be 1027 kg m^{-3} and 423 kg m^{-3} for CO_2 and CH_4 respectively (Ortiz Cancino et al. 2017). The error of the fitting process (Δ_n) was calculated through the standard deviation between experimental number of adsorbed moles and theoretical ones as in the Eq. (12.7).

$$\Delta_n = \frac{1}{N} \cdot \sqrt{\sum_{i=1}^{N} \left(n_{ads}^{exp} - n_{ads}^{FIT}\right)^2} \quad (12.7)$$

12.3 Results and Discussions

12.3.1 CH_4 and CO_2 Pure Gases Adsorption Isotherm

The Fig. 12.4 shows the results obtained for both CH_4 and CO_2 pure gases adsorption isotherms on CNR-115, CNR-115-ox and CNR-115-ox-am at 303.15 K and the pressure range of 0–3 MPa. The results highlight that, for the pressures ranging from zero to 3 MPa, activated carbons of CNR-115 family present higher adsorption capacities for CO_2 than for CH_4. Literately, this fact that carbon dioxide is highly adsorbed than methane, agrees with general behavior of activated carbons.

Further possible explanations are based on the difference between CO_2 and CH_4 properties (Table 12.2). From the Table 12.2 it can be seen that; the kinetic diameter of carbon dioxide (330 pm) is less than that of methane (380 pm). When compared to the size of activated carbons of CNR-115 family as previously presented, carbon dioxide (i.e. the one with small size) will tend to be adsorbed than methane (Cui et al. 2004). However, another reason is that, carbon dioxide doesn't present a dipole moment, but it contains unignorable quadrupole moment (-13.7×10^{40} cm^2), this is contrary to methane which has no dipole or quadrupole moments. The higher the dipole and quadrupole moments, the greater the interactions gas–activated carbon (Rodriguez-Reinoso et al. 1992). In addition, a wider polarizability of CO_2 than CH_4 (29.1×10^{25} cm^2 of CO_2 >25.9×10^{25} cm^3 of CH_4) also explains the greatest interactions between CO_2 and activated carbons. Besides, the results also show a significant loss in adsorption capacity for both CH_4 and CO_2 from commercial CNR-115 to modified activated carbons. At 3Mpa, CO_2 adsorption capacity was in the following sequence: CNR-115 (12.05 mol kg^{-1}) > CNR-115-ox (7.79 mol kg^{-1}) > CNR-115-ox-am (4.05 mol kg^{-1}), and was the same order for CH_4 adsorption capacity: CNR-115 (5.18 mol kg^{-1}) > CNR-115-ox (3.16 mol kg^{-1}) > CNR-115-ox-am (1.53 mol kg^{-1}). The highest CH_4 and CO_2 adsorption capacities of CNR-115 were surely related to its higher BET surface and higher pore volume. Those results were considered as accurate since the errors were very small ($\Delta_{nads} \leq 1\%$, $\Delta_T \leq 0.5\%$, and $\Delta_P = 1–5\%$).

12.3.2 Langmuir Fitting

The Table 12.3 shows the results obtained for the fitting of Langmuir model with CH_4 and CO_2 adsorption isotherms. The maximum Langmuir capacities of (8.45

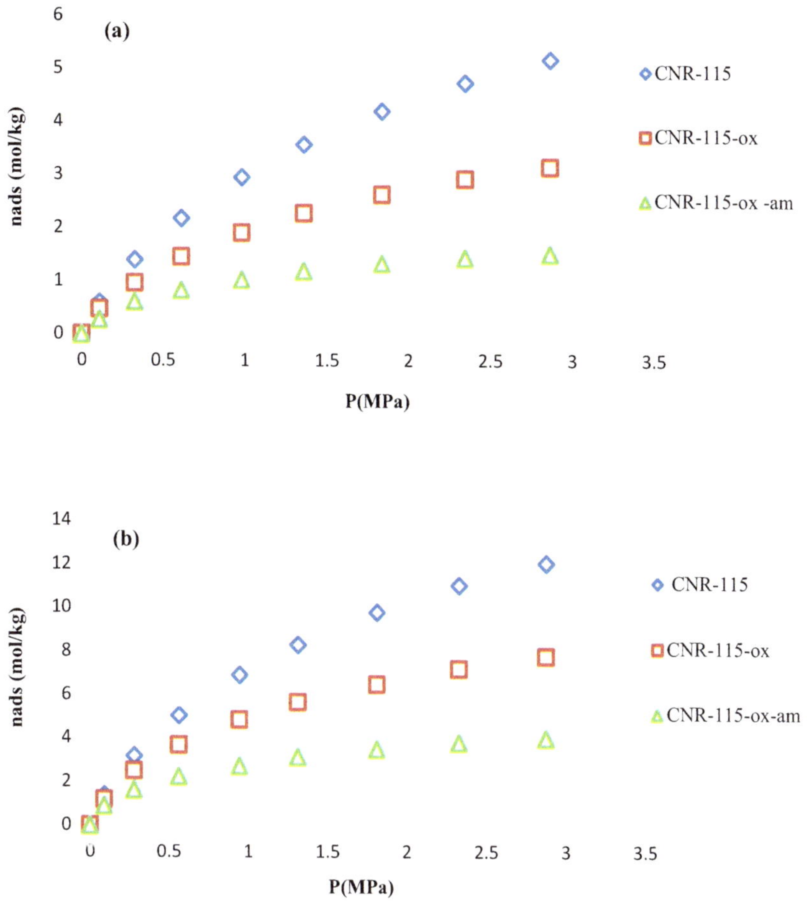

Fig. 12.4 CH_4 (**a**) and CO_2 (**b**) pure gases adsorption isotherms on CNR-115, CNR-115-ox and CNR-115-ox-am at 303.15 K

Table 12.2 Properties of CH_4 and CO_2 molecules (Wu et al. 2015)

Property	CH_4	CO_2
Kinetic diameter (pm)	380	330
Dipole moment ($\times 10^{-40}$ cm^2)	0	0
Quadrupole moment ($\times 10^{-40}$ cm^2)	0	−13.7
Polarizability ($\times 10^{-25}$ cm^3)	25.9	29.1

and 19.26 mol kg^{-1} for CH_4 and CO_2 respectively) were found to be for unmodified activated carbon (CNR-115). This confirms the relationship between the adsorption capacity and the BET surface area (i.e. the greater the BET surface the higher the adsorption capacity).

It can also be observed in Fig. 12.5 that, the maximum Langmuir capacity was in good correlation with BET surface area, with good linear regression ($R_2 \geq 0.98$). However, the results were confirmed as reasonable and reproductive since the fitting error was small ($\Delta_n \leq 0.12$) for all the samples.

Table 12.3 Langmuir parameters of CH_4 and CO_2 on activated carbons of CNR-115 family at 303.15 K

Sample	Components	n_L (mol kg^{-1})	p_L (MPa)	Δ_n
CNR-115	CH_4	8.45	1.79	0.003
	CO_2	19.26	1.68	0.073
CNR-115-ox	CH_4	4.78	1.53	0.040
	CO_2	9.71	0.94	0.120
CNR-115-ox-am	CH_4	2.10	1.04	0.030
	CO_2	4.39	0.57	0.010

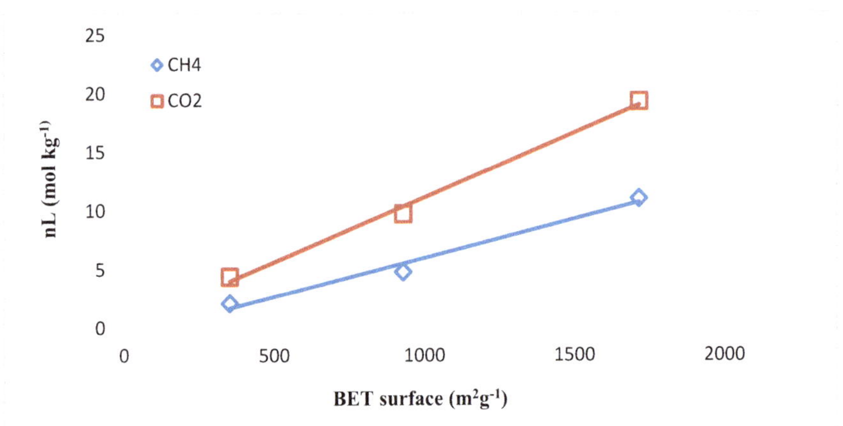

Fig. 12.5 Maximum Langmuir capacity versus BET surface area of activated carbons of the CNR-115 family, at 303.15 K

12.3.3 Interesting Side of CNR-115-ox-am: The New Modified Activated Carbon

The results discussed above highlight that CNR-115 adsorbs higher amount of both CH_4 and CO_2 compared to new modified activated carbons (CNR-115-ox and CNR-115-ox-am). However, the adsorption capacity is not the only one performance indicator of an adsorbent. The best one is therefore its selectivity (i.e. capacity to choose one component over another one from the mixture). For CH_4/CO_2 mixtures separation, an activated carbon which adsorbs the least CH_4 and moderated CO_2 can be preferred. This is because, CH_4 as the only one valuable component has to remain in the mixture for further uses (Costa et al. 2020). In this study, the lowest CH_4 adsorption capacity (1.53 mmol/g) was obtained for CNR-115-ox-am, thus the reason to consider it as promising adsorbent for CH_4 enrichment from CH_4/CO_2 mixtures. This was already confirmed by the results of Deneb et al. (Peredo-Mancilla et al. 2018) who studied the adsorption of (CH_4/CO_2 50:50) mixture on CNR-115, CNR-115-ox and CNR-115-ox-am. According to their results, the selectivity of unmodified CNR-115 (<2.7 through the whole pressure range) was almost 50 times less than the maximum selectivity value (129.0 at 2.17 MPa) for CNR-115-ox-am. Furthermore, when compared to the adsorption capacities from the literature as shown in Table 12.4, methane adsorption capacity obtained in this work for CNR-115-ox-am at 303.15 K and 0.1 MPa (0.27 mol/g), is to the best of our knowledge among the lowest ones that are so far reported for activated carbons. Moreover, adsorption performances of CNR-115-ox-am are comparable to the ones reported for metal organic frameworks like (Cu-MOF) (Wu et al. 2014).

12.4 Conclusion

In this experimental study, CH_4 and CO_2 pure gases adsorption experiments were correctly conducted on commercial and modified activated carbons (family of CNR-115) at 303.15 K and the pressure ranging from zero to three mega Pascals. Adsorption capacity of CO_2 was found to be higher than that of CH_4 for all the three activated carbons of CNR-115 family. A decrease in adsorption capacity was observed for both gases due to the decline in BET surface area and pore blockage which take place during modification process. Experimental isotherms were fitted to Langmuir two-parameter model in order to describe the behavior of adsorption process. Type I isotherm (i.e. monolayer adsorption) was therefore observed for CH_4 and CO_2 adsorption on a set of activated carbon of CNR-115 family. BET surface area was confirmed to be related to the adsorption capacity, and the values obtained are in good agreement with the ones reported in the literature (Peredo-Mancilla et al. 2018; Ho et al. 2021). The modified activated carbon CNR-115-ox-am presents the advantage of adsorbing the least methane than other activated carbons of the CNR-115 family,

Table 12.4 CH_4 and CO_2 pure gases adsorption on different adsorbents, at 0.1 MPa

Adsorbents	Modification process	BET (m^2/g)	T (K)	CH_4 uptake (mol/kg)	CO_2 uptake (mol/kg)	Ref.
CNR-115	–	1714	303	0.60	1.38	This work
CNR-115-ox	Air, 400 °C	929	303	0.47	1.18	This work
CNR-115-ox-am	NH_4OH, 24 h at 100 °C	352	303	0.27	0.91	This work
GAC	–	865	273	–	4.66	Giraldo et al. (2020)
GACO	HNO_3, 6 hat 100 °C	787	273	–	5.38	Giraldo et al. (2020)
GACA	NH_4OH, 24 h at 100 °C	634	273	–	7.57	Giraldo et al. (2020)
Pitch-based	–	1457	303	1.10	2.50	Wu et al. (2015)
Cu-MOF	–	105	298	0.47	0.86	Wu et al. (2014)
Cu-MOF	–	105	318	0.35	0.65	Wu et al. (2014)
MOF-1C	–	–	298	0.30	1.50	Bae et al. (2011)

which is thus the reason to consider this activated carbon as promising adsorbent for the future studies of CH_4 enrichment from CH_4/CO_2 mixtures.

Acknowledgements The author would like to acknowledge the framework agreement between BIOCAD E_2S of Université de Pau et des Pays de l'Adour (UPPA), CATLP (Communauté d'Agglomeration Tarbes-Lourdes-Pyrénées), the French program ANR (ANR-16-IDEX-0002) and CARNOT ISIFOR for their financial support.

References

Bae Y-S, Hauser BG, Farha OK, Hupp JT, Snurr RQ (2011) Enhancement of CO_2/CH_4 selectivity in metal-organic frameworks containing lithium cations. Microporous Mesoporous Mater 141:231–235

Bessières D, Lafitte T, Daridon J-L, Randzio SL (2005) High pressure thermal expansion of gases: measurements and calibration. Thermochim Acta 428:25–30

Boulinguiez B, Le Cloirec P (2015) Purification de biogaz - Élimination des COV et des siloxanes. Chim Verte

Cui A, Bustin RM, Dipple G (2004) Selective transport of CO_2, CH_4, and N_2 in coals: insights from modeling of experimental gas adsorption data. Fuel 83:293–303

da Costa RBR, Valle RM, Hernández JJ, Malaquias ACT, Coronado CJR, Pujatti FJP (2020) Experimental investigation on the potential of biogas/ethanol dual-fuel spark-ignition engine for power generation: combustion, performance and pollutant emission analysis. Appl Energy 261:114438

El-Shafey EI, Ali SNF, Al-Busafi S, Al-Lawati HAJ (2016) Preparation and characterization of surface functionalized activated carbons from date palm leaflets and application for methylene blue removal. J Environ Chem Eng 4:2713–2724
Gasquet V (2020) Epuration d'H2S du biogaz à partir de résidus de traitement thermique bruts et formulés: Comparaison des performances et compréhension des mécanismes d'adsorption 223
Giraldo L, Vargas DP, Moreno-Piraján JC (2020) Study of CO_2 adsorption on chemically modified activated carbon with nitric acid and ammonium aqueous. Front Chem 8:543452
Ho BN, Pino-Perez D, Matei Ghimbeu C, Diaz J, Peredo-Mancilla D, Hort C, Bessieres D (2021) Determination of methane, ethane and propane on activated carbons by experimental pressure swing adsorption method. J Nat Gas Sci Eng 95:104124
Jagiello J, Olivier JP (2013) 2D-NLDFT adsorption models for carbon slit-shaped pores with surface energetical heterogeneity and geometrical corrugation. Carbon Complete 70–80
Kapoor R, Ghosh P, Kumar M, Vijay VK (2019) Evaluation of biogas upgrading technologies and future perspectives: a review. Environ Sci Pollut Res Int 26:11631–11661
Madzaki H, KarimGhani WAWAB, Rebitanim NZ, Alias AB (2016) Carbon dioxide adsorption on sawdust biochar. Procedia Eng 148:718–725
Mouahid A, Bessieres D, Plantier F and Pijaudier-Cabot G (2011) A thermostated coupled apparatus for the simultaneous determination of adsorption isotherms and differential enthalpies of adsorption at high pressure and high temperature. J Therm Anal Calorim
Olivier T (2015) Étude technique, Économique et environnementale sur l'injection portée de biomethane dans le réseau de gaz
Ortiz Cancino OP, Pino Pérez D, Pozo M, Bessieres D (2017) Adsorption of pure CO_2 and a CO_2/CH_4 mixture on a black shale sample: manometry and microcalorimetry measurements. J Pet Sci Eng 159:307–313
Peredo Mancilla JD (2019) Adsorption and separation of carbon dioxide for biomethane production: the use of activated carbons. Thèse de doctorat (Pau)
Peredo-Mancilla D, Hort C, Jeguirim M, Ghimbeu CM, Limousy L, Bessieres D (2018) Experimental determination of the CH_4 and CO_2 pure gas adsorption isotherms on different activated carbons. J Chem Eng Data 63:3027–3034
Pu Q, Zou J, Wang J, Lu S, Ning P, Huang L, Wang Q (2021) Systematic study of dynamic CO_2 adsorption on activated carbons derived from different biomass. J Alloys Compd 887:161406
Ribeiro R, Sauer T, Lopes F, Moreira R, Grande C, Rodrigues A (2008 Adsorption of CO_2, CH_4, and N_2 in activated carbon honeycomb monolith. J Chem Eng Data 53:2311–2317
Rodriguez-Reinoso F, Molina-Sabio M, Munecas MA (1992) Effect of microporosity and oxygen surface groups of activated carbon in the adsorption of molecules of different polarity. J Phys Chem 96:2707–2713
Rogulska M, Bukrejewski P, Krasuska E (2018) Biomethane as transport fuel (IntechOpen)
Speight JG (2019) Chapter 15—Hydrogen production. In: Speight JG (ed) Heavy oil recovery and upgrading, pp 657–697. Gulf Professional Publishing
Sun L-M, Meunier F, Brodu N and Manero M-H (2016) Adsorption—aspects théoriques
Tilagone R, Lecointe B (2015) Gaz naturel—Carburant pour véhicule 31
Vondra M, Touš M, Teng SY (2019) Digestate evaporation treatment in biogas plants: a techno-economic assessment by Monte Carlo, neural networks and decision trees. J Clean Prod 238:117870
Wu X, Yuan B, Bao Z, Deng S (2014) Adsorption of carbon dioxide, methane and nitrogen on an ultramicroporous copper metal–organic framework. J Colloid Interface Sci 430:78–84
Wu Y-J, Yang Y, Kong X-M, Li P, Yu J-G, Ribeiro AM, Rodrigues AE (2015) Adsorption of pure and binary CO_2, CH_4, and N_2 gas components on activated carbon beads. J Chem Eng Data 60:2684–2693
Zulkurnai NZ, Md Ali UF, Ibrahim N, Abdul Manan NS (2017) Carbon dioxide (CO_2) adsorption by activated carbon functionalized with deep eutectic solvent (DES). IOP Conf Ser Mater Sci Eng 206:012001

Chapter 13
Optimization of Dual-Chamber Microbial Fuel Cells for the Biodegradation of Acetaminophen

I. S. Asetre and L. L. Tayo

Abstract Microbial fuel cells (MFCs) are emerging hybrid bioreactors that offers renewable energy source. MFCs are promising technologies that generates bioelectrical energy using different kinds of microorganisms to oxidize a various types of fuel substrates and simultaneously works as a wastewater treatment unit. This study evaluated the performance of MFCs, inoculated with active soil microorganism and crabmeat processing wastewater as fuel, in the degradation of a pharmaceutical compound acetaminophen, while generating bioelectricity. Acetaminophen quantification was carried out using high-performance liquid chromatography with photodiode array (HPLC-PDA) detector. At an initial 200 mg/L concentration, in MFC-GTS and MFC-Nafion with 1:3 ratio of microbes to wastewater, using carbon brush electrodes, recorded removal percentage of 26.75% and 20.5%, respectively. This revealed that MFCs with active soil microbes is a promising approach in the degradation of acetaminophen in wastewater. Further, MFCs inoculated with active indigenous soil microorganisms and fed with crab meat processing wastewater is a potential source of bioelectricity.

Keywords Microbial fuel cells · Acetaminophen · Degradation

I. S. Asetre
Environmental Science – Tinambac Campus, Partido State University, 4422 Camarines Sur, Philippines

L. L. Tayo (✉)
School of Chemical, Biological and Materials Engineering and Sciences, Mapua University, 1002 Manila, Philippines
e-mail: lltayo@mapua.edu.ph

J. Baeyens et al. (eds.), *Proceedings of 2022 4th International Conference on Environment Sciences and Renewable Energy*, Environmental Science and Engineering, https://doi.org/10.1007/978-981-19-9440-1_13

13.1 Introduction

MFCs, defined as promising bio-electrochemical hybrid biorectors (Palanisamy et al. 2019), simple, robust technology (Rabaey and Verstraete 2005), are known to using bacteria to break down organic matters thru catalytic activities under anaerobic conditions (Du et al. 2007) where there is no input of combined heat and power unit (Ye et al. 2019). MFCs are novel approach for the sustainable generation of energy from these biodegradable, reduced compounds (Rabaey and Verstraete 2005) and perform simultaneously as wastewater treatment units (Cheng et al. 2021; Saba et al. 2021; Al-Ansari et al. 2021; Zhu et al. 2021; Sathe et al. 2021; Gul et al. 2021; Sani et al. 2021; Verma et al. 2021). MFC system as a wastewater treatment unit has drawn great interests in the recent years. Many healthy investigations for microbial fuel cell on the remediation surface water (Wang et al. 2019), soils and groundwater (Guan et al. 2019; Habibul et al. 2016; Zhao et al. 2019; Zhang et al. 2019), of river sediments (Kabutey et al. 2019; Li et al. 2017; Nastro et al. 2019) and reclamation of polluted waters have been conducted by many researchers to promote the integrated applications of this technology.

On the other hand, pharmaceuticals and personal care products or PPCPs are considered contaminants of emerging concern due to their suspected and known impacts to the environment and risks to human health which are potentially hazardous even at low concentrations. Over the past decades, overwhelming evidence has shown that PPCPs are ubiquitous in surface waters (Liu et al. 2020; Yuan et al. 2020; Ebele et al. 2017; Katsikaros and Chrysikopoulos 2021; Li et al. 2021), groundwater (Khan et al. 2022; Ebele et al. 2020), and detected in drinking water (Cao et al. 2020; Pai et al. 2020) even in wastewater and sewage treatment facilities (Clyde et al. 2021; Ma et al. 2022; Singh and Suthar 2021; Al-Baldawi et al. 2021; Hena et al. 2021). These detections reflected inputs from both point and non-point sources. One of the pharmaceutical compounds frequently detected in an aquatic environment is the acetaminophen, belonging to the drug classification of PPCPs. Also known as paracetamol, acetaminophen is considered as an endocrine disrupting (EDC) compound. When emitted into the environment, this compound alters the physiological processes of and is toxic to aquatic organisms. Acetaminophen has shown to impair and malform the development of catfish embryos exposed to environmental concentrations (μg/L) and had an acute toxicity in catfish early-life stages (Erhunmwunse et al. 2021). Rainbow trout exposed to 0, 10 and 30 μg/L acetaminophen for 4 weeks significantly impacted kidney, gill and liver function, as well as critical swimming speed (Choi et al. 2018). Chronic exposure to acetaminophen affected the survival of *Oryzias latipes* at levels much higher than environment while prenatal exposure to acetaminophen reduced hatching success of its eggs (Kim et al. 2012). Another study reported the altered liver histology, increased developmental abnormalities and mortality as the effect of chronic exposure of zebrafish *Danio rerio* to environmentally relevant concentrations of paracetamol or acetaminophen (Galus et al. 2013).

Many works have been published and available online on the removal of PPCPs in aquatic environment and waste stream utilizing varied materials and technologies, however, studies on the use of MFCs in the removal of acetaminophen remain very scarce, despite the numerous data on the detection in different ecological systems. Most studies thus far have focused on the presence in maximum and minimum level concentrations of these compounds in the environment and its significant impacts on the aquatic organisms.

This study presented and discussed the optimised performance of active soil microbes-inoculated MFCs to degrade acetaminophen in different concentrations of anolytes, using different proton exchange membranes and electrode materials, while simultaneously generating bioelectricity. Crab meat processing wastewater was used as fuel substrate. This study confirms that MFCs installed with GTS and inoculated with active soil microorganisms is effective in the degradation of acetaminophen in wastewater. The removal efficiency of the MFC systems is comparable to other studies utilizing other technologies. The results of this study may provide relevant information in the promotion of MFC green technology to provide an alternative approach in the reclamation of acetaminophen-contaminated wastewater facilitating microbial degradation by active soil microorganisms while simultaneously generating electricity.

13.2 Materials and Methods

13.2.1 Reactor Architecture and Operations

The MFC bioreactors were made of modified Pyrex reagent bottles (from BE Scientific Glass Instruments), with anode and cathode chambers separated by membranes. The bioreactors has an operational volume of 250 mL per chamber, air space of approximately 17%, and side ports for the injection and discharge of anolytes, catholytes and the pollutant. The anode and cathode chambers are separated by a specific proton exchange membranes. For MFC-Nafion system, a 50.8 μm thick Dupont™ Nafion 212® membrane acquired from FuelCells, Etc. was used. The nafion sheet was cut into square 4 × 4 cm, and activated according published studies (Fan and Miao 2015; Angioni et al. 2016; Asetre and Tayo 2021; Bartolome and Tayo 2020). For MFC-GTS system, a modified gelatinized tapioca starch membrane was utilised. The starch extraction and gelatinization process was patterned to the methods in published studies (Daramola and Aina 2007; Loh et al. 2008; Obasi et al. 2012) with slight modification. All reactors were fixed with rubber ring, and clamps similar to the MFC architecture and set-up of the published studies (Asetre and Tayo 2021; Bartolome and Tayo 2020). The titanium mesh wire electrodes used in the experiments are bent to form a cylinder (Asetre and Tayo 2021; Bartolome and Tayo 2020). The carbon fiber brush electrodes are made from PANEX 35 carbon-filled materials twisted to titanium core (from Mill-Rose Company) and were acid and heat treated following procedures defined in published literatures (Asetre and Tayo 2021; Bartolome and Tayo 2020; Feng et al. 2010; Higgins et al. 2013; Ma et al. 2016). The brushes and mesh wires were repeatedly rinsed in deionized water before using. During the experiments, all MFC systems were fed with different concentrations of anolyte. The anolyte are of different ratios of microbial inoculant to wastewater substrate which is 1:0, 1:1, 1:3, 0:1. The cathode chamber is fed with 0.01 KMNO4 solution and continuously aerated throughout the monitoring. The MFCs were run in batch operations and in closed circuit condition; outputs were computed after determining the optimum running conditions and the stabilised open circuit potential (OCP) values. The cathodes were coupled to anode with a consistent 200-ohm external load connection. Acetaminophen BioXtra, $\geq$99.0% (Merck) was used in the pollutant injection set-up. The current profile of the MFC systems with acetaminophen was recorded during the pollutant injection experiments. 10 mL of samples were withdrawn from the MFC reactors after 21 h of continuous operation and was submitted to Bicol University, Legaspi City, Albay, Camarines Sur for HPLC-PDA analysis.

Collection, Cultivation and Stabilization of the Soil Microorganisms. The collection of microbes employed similar methods with published studies (Asetre and Tayo 2021; Anwanyu et al. 2015), where cooked organic rice was set loosely in a wooden box, covered with sterile paper towel and tissue and placed in undisturbed bamboo litter for 3 days. The sample was collected after a mycelium has grown into mass, fermented for 15 days in a muscovado sugar in 1:1 ratio (Park and DuPonte 2010) and stabilized by adding three parts of non-chlorinated water. The stabilised sample is stored in an opaque container and stored in a cool, dry place.

Simulation and Characterization of Wastewater and Data Collection. To give objectivity in pollutant injection experiments comprising crab meat processing wastewater, the process wastewater from the local small scale crab meat processing plant was simulated. Similar process of wastewater simulation was employed (Asetre and Tayo 2021) where 10 kilos of live market size blue crabs were cooked in a steam of 4L of water. The wastewater is obtained from the combined liquids from washing of raw crabs, steaming process, and washing of crabmeat. The wastewater was analysed for temperature, pH, TDS, electrical conductivity using Hach® sensIon5 meter and General handheld instruments. Hach® COD kit Method 8000 was used to measure COD. The physico-chemical characteristics of the simulated wastewater are as follows: pH—5.27, COD—6,143 ppm, total dissolved solids—2,530 ppm, turbidity—67 FNU, electrical conductance—4,870 μS/cm, salinity—2.6 ppt, temperature—29–30 °C. High COD wastewater to use as a substrate in MFCs is considered ideal for higher power density output.

For the data collection, a voltage data collection system (from Layad Circuits Electronics Engineering Supplies and Services) was used for the monitoring and recording of the cell potential of the MFCs, recording outputs every 30 s. Digital multimeters were connected to the MFCs to validate the data collected by the voltage data system. The recorded outputs were computed for power (P), power density $P_{density}$, and the energy (E) values, using the equations:

$$P = I * V \tag{13.1}$$

$$P_{density} = P/A \tag{13.2}$$

$$CCP(I) = V/R \tag{13.3}$$

where: I = V/R (Ohm's Law), V = voltage output, R = fixed external resistance of 200 Ώ, A = surface area.

13.3 Results and Discussion

13.3.1 Open Circuit Potential of MFCs

The cell potential of the MFCs with 0:1 anolyte were measured after the inoculant stabilization, where active soil microbes biofilm has attached in the electrodes. In a separate study (Asetre and Tayo 2021), the MFC-Nafion with titanium mesh electrodes generated peak output 967 mV from the initial values of 801 mV, while using carbon brush electrodes generated lower peak value of 829 mV from initial value of 702 mV. On the other hand, MFC-GTS using titanium mesh electrodes generated peak output of 960 mV from initial value of 869 mV, and using carbon brush electrodes generated lower peak value of 858 mV from initial output of 721 mV.

Closed Circuit Current Profile of MFCs. In MFC-Nafion using titanium wire electrodes, the average current values recorded was 2.80 mA for 1:1 anolyte, 2.85 mA for 1:3 and 2.59 mA both for 0:1 and 1:0. The current generated was higher in 1:3 anolyte which peaked at 2.98 mA. Using carbon brush electrodes, the average current values recorded was 2.95 mA, 3.01 mA, 2.80 mA, and 2.88 mA for 1:1, 1:3, 0.1, 1:0, anolytes, respectively. The current generated remains higher in 1:3 anolyte which peaked at 3.20 mA. On the other hand, in MFC-GTS using titanium wire electrodes recorded average values of 3.17 mA, 3.19 mA, 2.96 mA, and 2.92 mA for 1:1, 1:3, 0:1, 1:0, anolytes, respectively. Carbon brush electrodes recorded average current values of 3.42 mA, 3.52 mA, 3.13, mA, and 3.15 mA for the same anolytes. Similarly, the current generation remains higher in 1:3 anolyte which peaked at 3.35 mA and 3.62 mA using titanium wire and carbon brush electrodes, respectively. The current in all MFC set-up shows that MFC-GTS with carbon brush electrodes produced higher output. For both MFC set-up, the increase in the average current output is proportional to the volume of wastewater. When the wastewater in the anolyte increases, more food is available for the bacteria to consume, thereby generating higher current output.

Volumetric Power Density. The computed average power density calculated from the volumetric surface area of MFC-GTS is compared to previous studies (Pandey et al. 2016). Figure 13.1 shows the average power density generated by MFC-GTS with 1:3 anolyte using titanium mesh electrode (9,908 mW/m^3) and carbon brush

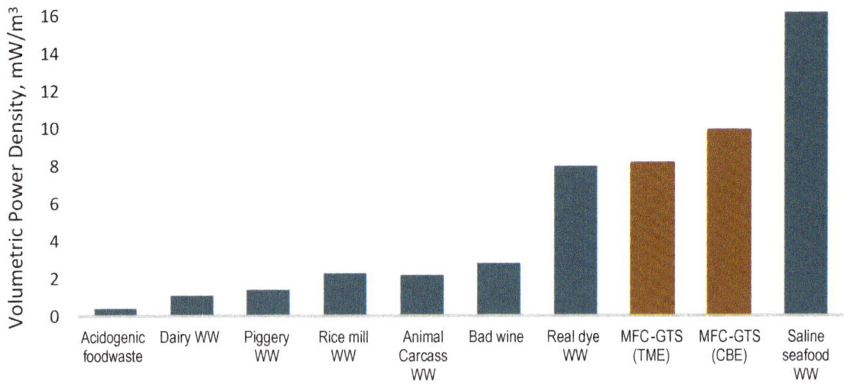

Fig. 13.1 Comparison of the maximum volumetric power density values (in mW/m^3) of MFCs using different substrates

electrode (8,158 mW/m^3). The performance of the two MFC systems used in this study revealed higher power density output than the previous studies using different substrate.

Acetaminophen Degradation in MFC-Nafion Reactors. Presented in Fig. 13.2 is the degradation of acetaminophen in MFC-Nafion bioreactor using different electrode materials showing initial (Ci) and final (Cf) concentrations. Using titanium mesh electrodes with 1:3 anolyte, MFC-Nafion recorded 12% removal with final concentration of 176 mg/L, followed by 10% for 1:1 anolyte with 180 mg/L final concentration, with the same operating conditions.

Using carbon fiber brush with 1:3 anolyte, the recorded removal percentage is 20.5% with the final concentration of 159 mg/L, while MFC-Nafion with 1:1 anolyte recorded 17% removal, with the same operating conditions.

Acetaminophen Degradation in MFC-GTS Reactors. Presented in Fig. 13.3 is the degradation of acetaminophen in MFC-GTS reactor using different electrode

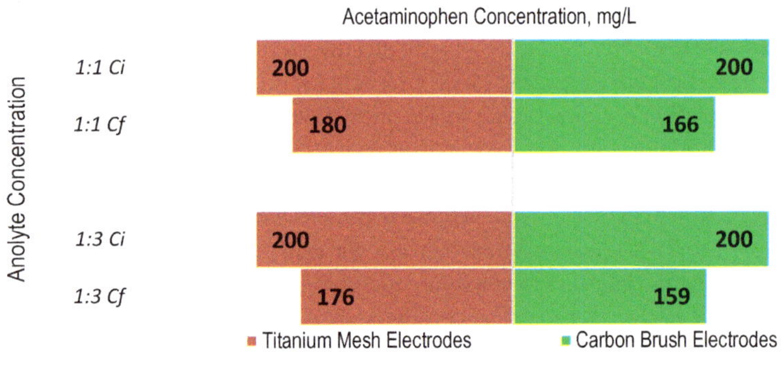

Fig. 13.2 Degradation of acetaminophen in MFC-Nafion using different electrodes

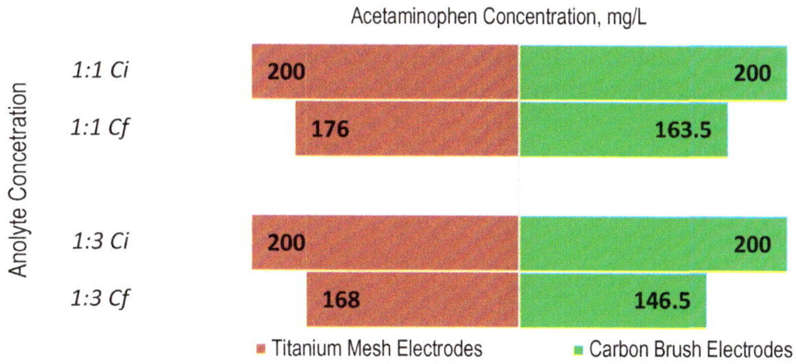

Fig. 13.3 Degradation of acetaminophen in MFC-GTS using different electrodes

materials. MFC-GTS using titanium wire mesh electrodes fed with 1:3 anolyte also recorded 16.3% removal, followed by 13.3% for 1:1 anolyte with the same operating conditions. MFC-GTS using carbon fiber brush fed with 1:3 anolyte recorded 26.5% removal percentage with final acetaminophen concentration of 146.5 mg/L. The system fed by 1:1 anolyte recorded 18.3% removal percentage with final concentration of 163 mg/L, with the same operating conditions.

At an initial experimental concentration of 200 mg/L, acetaminophen dropped at 146.5 mg/L in MFC-GTS with 1:3 ratio of anolyte using carbon brush electrodes (with removal percentage 26.75% or 2.55 mg/L per hour) after 21 h of continuous operations. This removal percentage value was highest among all experiment setup. These results were comparable to a study (López Zavala and Estrada 2016) using innovative electrochemical oxidation cell where 10 mg/L concentration of acetaminophen in solution which achieved total degradation in a reaction time of 240 min (or 2.5 mg/L per hour removal).

13.4 Conclusion and Recommendations

This study confirms that MFCs inoculated with indigenous soil microbes could be a viable technological solution for the biodegradation of acetaminophen in wastewater. After 21 h of operation, acetaminophen concentrations significantly reduced. The removal efficiency was comparable to other studies utilizing other technologies. Further, MFCs using carbon fiber brushes are consistent in producing a more stable and higher current output. MFC-GTS achieved higher and more stable current than MFC-Nafion. Anolytes containing 25% active soil microbes (1:3) produced high current output achieved higher acetaminophen removal in the systems. Results of the experiments also confirms that active soil microorganisms can be used as a practical biocatalyst in MFCs. The use of gelatinized tapioca starch as membrane and active soil microbes as microbial catalyst in MFCs offers a relatively low cost and easily accessible MFC components. This study can create a better perspective on the other uses of MFC green technology in the Philippines focusing on the energy sustainability for wastewater treatment plants and its integration into wastewater infrastructure to meet effluent quality set by regulatory bodies while considering operational cost.

Acknowledgments This work was supported by the Department of Science and Technology—Engineering Research and Development for Technology (ERDT) research grant.

References

Al-Ansari MM, Benabdelkamel H, Al-Humaid L (2021) Degradation of sulfadiazine and electricity generation from wastewater using Bacillus subtilis EL06 integrated with an open circuit system. Chemosphere 276:130145

Al-Baldawi IA, Mohammed AA, Mutar ZH, Abdullah SRS, Jasim SS, Almansoory AF, Ismail NI (2021) Application of phytotechnology in alleviating pharmaceuticals and personal care products (PPCPs) in wastewater: source, impacts, treatment, mechanisms, fate, and SWOT analysis. J Clean Prod 319:128584

Angioni S, Millia S, Bruni G, Tealdi C, Mustarelli P, Quartarone E (2016) Improving the performances of Nafion^{TM}-based membranes for microbial fuel cells with silica-based, organically-functionalized mesostructured fillers. J Power Sources 334:120–127

Anwanyu CF, Ngohayon SL, Ildefonso RL, Ngohayon JL (2015) Isolation and characterization of indigenous microorganism (Imo) from Ifugao bamboo (Phyllostachys aurea) forest. Int J Sci Res 4(2):1319–1324

Asetre IS, Tayo LL (2021) Performance of a dual-chamber microbial fuel cell inoculated with active soil microbes for bioelectricity generation. In: 10th IEEE International conference on renewable energy research and applications, pp 159–164

Bartolome GJC, Tayo LL (2020) Decolorization of methyl orange using a double-chamber microbial fuel cell with the use of soil microorganisms. IOP Conf Ser Earth Environ Sci 563(1)

Cao SS, Duan YP, Tu YJ, Tang Y, Liu J, Zhi WD, Dai C (2020) Pharmaceuticals and personal care products in a drinking water resource of Yangtze River Delta Ecology and Greenery Integration Development Demonstration Zone in China: occurrence and human health risk assessment. Sci Total Environ 721:137624

Cheng D, Ngo HH, Guo W, Chang SW, Nguyan DD, Liu Y, Liu Y, Deng L, Chen Z (2021) Evaluation of a continuous flow microbial fuel cell for treating synthetic swine wastewater containing antibiotics. Sci Total Environ 756:144133

Choi E, Alsop D, Wilson JY (2018) The effects of chronic acetaminophen exposure on the kidney, gill and liver in rainbow trout (Oncorhynchus mykiss). Aquat Toxicol 198:20–29

Clyde PM, Lee CS, Price RE, Venkatesan AK, Brownawell BJ (2021) Occurrence and removal of PPCPs from on-site wastewater using nitrogen removing biofilters. Water Res 206:117743

Daramola B, Aina JO (2007) Effects of alum on pasting and some physicochemical properties of cassava (Manihot esculentum) starch. World J Diary Food Sci 2(1):18–22

Du Z, Li H, Gu T (2007) A state of the art review on microbial fuel cells: a promising technology for wastewater treatment and bioenergy. Biotechnol Adv 25(5):464–482

Ebele AJ, Abou-Elwafa Abdallah M, Harrad S (2017) Pharmaceuticals and personal care products (PPCPs) in the freshwater aquatic environment. Emerg Contam 3(1):1–16

Ebele AJ, Oluseyi T, Drage DS, Harrad S, Abou-Elwafa Abdallah M (2020) Occurrence, seasonal variation and human exposure to pharmaceuticals and personal care products in surface water, groundwater and drinking water in Lagos State, Nigeria. Emerg Contam 6:124–132

Erhunmwunse NO, Tongo I, Ezemonye LI (2021) Acute effects of acetaminophen on the developmental, swimming performance and cardiovascular activities of the African catfish embryos/larvae (Clarias gariepinus). Ecotoxicol Environ Saf 208:111482

Fan L, Miao X (2015) Performance study on microbial fuel cell treating restaurant wastewater. Nat Environ Pollut Technol 14(3):691

Feng Y, Yang Q, Wang X, Logan BE (2010) Treatment of carbon fiber brush anodes for improving power generation in air-cathode microbial fuel cells. J Power Sources 195(7):1841–1844

Galus M, Kirischian N, Higgins S, Purdy J, Chow J, Rangaranjan S, Li H, Metcalfe C, Wilson JY (2013) Chronic, low concentration exposure to pharmaceuticals impacts multiple organ systems in zebrafish. Aquat Toxicol 132–133:200–211

Guan CY, Tseng YH, Tsang DCW, Hu A, Yu CP (2019) Wetland plant microbial fuel cells for remediation of hexavalent chromium contaminated soils and electricity production. J Hazard Mater 365:137–145

Gul H, Raza W, Lee J, Azam M, Ashraf M, Kim KH (2021) Progress in microbial fuel cell technology for wastewater treatment and energy harvesting. Chemosphere 281:130828

Habibul N, Hu Y, Sheng GP (2016) Microbial fuel cell driving electrokinetic remediation of toxic metal contaminated soils. J Hazard Mater 318:9–14

Hena S, Gutierrez L, Croué JP (2021) Removal of pharmaceutical and personal care products (PPCPs) from wastewater using microalgae: a review. J Hazard Mater 403:124041

Higgins SR, Lopez RJ, Pagaling E, Yan T, Cooney MJ (2013) Towards a hybrid anaerobic digester-microbial fuel cell integrated energy recovery system: an overview of the development of an electrogenic biofilm. Enzyme Microb Technol 52(6–7):344–351

Kabutey FT, Ding J, Zhao Q, Antwi P, Quashie FK, Tankapa V, Zhang W (2019) Pollutant removal and bioelectricity generation from urban river sediment using a macrophyte cathode sediment microbial fuel cell (mSMFC). Bioelectrochemistry 128:241–251

Katsikaros AG, Chrysikopoulos CV (2021) Occurrence and distribution of pharmaceuticals and personal care products (PPCPs) detected in lakes around the world—a review. Environ Adv 6:100131

Khan HK, Rehman MYA, Junaid M, Lv M, Yue L, Haq IU, Xu N, Malik RN (2022) Occurrence, source apportionment and potential risks of selected PPCPs in groundwater used as a source of drinking water from key urban-rural settings of Pakistan. Sci Total Environ 807:151010

Kim PG, Park Y, Ji K, Seo J, Lee S, Choi K, Kho Y, Park J, Choi K (2012) Effect of chronic exposure to acetaminophen and lincomycin on Japanese medaka (Oryzias latipes) and freshwater cladocerans Daphnia magna and Moina macrocopa, and potential mechanisms of endocrine disruption. Chemosphere 89(1):10–18

Li H, He W, Qu Y, Li C, Tian Y, Feng Y (2017) Pilot-scale benthic microbial electrochemical system (BMES) for the bioremediation of polluted river sediment. J Power Sources 356(73):430–437

Li L, Zhao X, Liu D, Song K, Liu Q, He Y (2021) Occurrence and ecological risk assessment of PPCPs in typical inflow rivers of Taihu lake, China. J Environ Manag 285:112176

Liu N, Jin X, Feng C, Wang Z, Wu F, Johnson AC, Xiao H, Hollert H, Giesy JP (2020) Ecological risk assessment of fifty pharmaceuticals and personal care products (PPCPs) in Chinese surface waters: a proposed multiple-level system. Environ Int 136(8):105454

Loh SL, Rajan A, Wong SC, Bong CFJ (2008) Physiochemical properties of sago starch modified by acid treatment in alcohol. 5(4) 307–311

López Zavala MA, Estrada EE (2016) Degradation of acetaminophen and its transformation products in aqueous solutions by using an electrochemical oxidation cell with stainless steel electrodes. Water 8:1–12

Ma J, Ni H, Su D, Meng X (2016) Bioelectricity generation from pig farm wastewater in microbial fuel cell using carbon brush as electrode. Int J Hydrogen Energy 41(36):16191–16195

Ma L, Liu Y, Yang Q, Jiang L, Li G (2022) Occurrence and distribution of pharmaceuticals and personal care products (PPCPs) in wastewater related riverbank groundwater. Sci Total Environ 821(59):153372

Nastro RA, Gambino E, Toscanesi M, Arienzo M, Ferrara L, Trifuoggi M (2019) Microbial fuel cells (MFCs) remediation activity of marine sediments sampled at a dismissed industrial site: what opportunities? J Clean Prod 235:1559–1566

Obasi LA, Opara CC, Oji A (2012) Performance of cassava starch as a proton exchange membrane in a dual chambered microbial fuel cell. Int J Eng Sci Technol 4(01):227–238

Pai CW, Leong D, Chen CY, Wang GS (2020) Occurrences of pharmaceuticals and personal care products in the drinking water of Taiwan and their removal in conventional water treatment processes. Chemosphere 256:127002

Palanisamy G, Jung HY, Sadhasivam T, Kurkuri MD, Kim SC, Roh SH (2019) A comprehensive review on microbial fuel cell technologies: processes, utilization, and advanced developments in electrodes and membranes. J Clean Prod 221:598–621

Pandey P, Shinde VN, Deopurkar RL, Kale SP, Patil SA, Pant D (2016) Recent advances in the use of different substrates in microbial fuel cells toward wastewater treatment and simultaneous energy recovery. Appl Energy 168:706–723

Park H, DuPonte MW (2010) How to cultivate indigenous microorganisms. Biotechnology 9:1–7

Rabaey K, Verstraete W (2005) Microbial fuel cells: novel biotechnology for energy generation. Trends Biotechnol 23(6):291–298

Saba B, Kjellerup BV, Christy AD (2021) Eco-friendly bio-electro-degradation of textile dyes wastewater. Bioresour Technol Rep 15:100734

Sani AM, Savla N, Pandit S, Mathuriya AS, Gupta PK, Khanna N, Babu RP, Kumar S (2021) Recent advances in bioelectricity generation through the simultaneous valorization of lignocellulosic biomass and wastewater treatment in microbial fuel cell. Sustain Energy Technol Assess 48:101572

Sathe SM, Chakraborty I, Sankar Cheela VR, Chowdhury S, Dubey BK, Ghangrekar MM (2021) A novel bio-electro-Fenton process for eliminating sodium dodecyl sulphate from wastewater using dual chamber microbial fuel cell. Bioresour Technol 341:125850

Singh V, Suthar S (2021) Occurrence, seasonal variation, mass loading and fate of pharmaceuticals and personal care products (PPCPs) in sewage treatment plants in cities of upper Ganges bank, India. J Water Process Eng 44:102399

Verma P, Daverey A, Kumar A, Arunachalam K (2021) Microbial fuel cell—a sustainable approach for simultaneous wastewater treatment and energy recovery. J Water Process Eng 40:101768

Wang H, Fu B, Xi J, Hu HY, Liang P, Huang X, Zhang X (2019) Remediation of simulated malodorous surface water by columnar air-cathode microbial fuel cells. Sci Total Environ 687:287–296

Ye Y, Ngo HH, Guo W, Chang SW, Nguyen DD, Liu Y, Nghiem LD, Zhang X, Wang J (2019) Effect of organic loading rate on the recovery of nutrients and energy in a dual-chamber microbial fuel cell. Bioresour Technol 281:367–373

Yuan X, Hu J, Li S, Yu M (2020) Occurrence, fate, and mass balance of selected pharmaceutical and personal care products (PPCPs) in an urbanized river. Environ Pollut 266:115340

Zhang T, Hu L, Zhang M, Jiang M, Fiedler H, Bai W, Wang X, Zhang D, Li Z (2019) Cr(VI) removal from soils and groundwater using an integrated adsorption and microbial fuel cell (A-MFC) technology. Environ Pollut

Zhao L, Deng J, Hou H, Li J, Yang Y (2019) Investigation of PAH and oil degradation along with electricity generation in soil using an enhanced plant-microbial fuel cell. J Clean Prod 221:678–683

Zhu CY, Wang JF, Li QS, Wang LL, Tang GH, Cui BS, Bai J (2021) Integration of CW-MFC and anaerobic granular sludge to explore the intensified ammonification-nitrification-denitrification processes for nitrogen removal. Chemosphere 278:130428

Chapter 14
Study on the Relationship Between Energy Consumption of Shipbuilding and Shipbuilding Costs

Yuhang Zhai

Abstract This paper firstly elaborated the background of green manufacturing in China's shipbuilding industry, and then analyzed the types of energy consumption that need to be invested in the implementation of green manufacturing in the shipbuilding industry based on the environmental value chain. Then, based on the panel data of shipbuilding industry energy consumption of provinces, municipalities and autonomous regions, the relationship between energy consumption and pollutant emission and shipbuilding cost changes in shipbuilding process was studied by using Ridge Regression model. It is found that changes in the amount of energy consumption and the amount of pollutant emissions in the shipbuilding industry will lead to changes in shipbuilding costs to varying degrees. The conclusion points out that in order to develop green economy, shipbuilding enterprises must actively reduce energy consumption in shipbuilding process and gradually promote green energy instead of traditional energy to achieve the purpose of energy saving and cost reduction.

Keywords Shipbuilding · Green economy · Energy consumption · Green energy resource · Ridge regression model

14.1 Introduction

With the acceleration of industrialization and urbanization in China in recent years, the gross value of China's shipbuilding industry accounts for an increasingly high proportion of the gross national product (Wang 2021). At the same time, it also brings about environmental problems that cannot be ignored. The production and manufacturing process of ships has inherent characteristics such as large volume and long cycle, and the energy consumption of ship industry occupies a large proportion in

Y. Zhai (✉)
School of Economics and Management, Jiangsu University of Science and Technology, Zhenjiang, Jiangsu Province, China
e-mail: zyh375606@163.com

J. Baeyens et al. (eds.), *Proceedings of 2022 4th International Conference on Environment Sciences and Renewable Energy*, Environmental Science and Engineering, https://doi.org/10.1007/978-981-19-9440-1_14

all industrial fields. The contradictory issue of economic development and environmental protection is the first problem in the field of industrial production. Since the 18th Party Congress, the Party Central Committee has attached great importance to environmental protection and written the construction of ecological civilization as a social development goal into the five-in-one layout policy, economic development cannot be at the expense of the environment.

At present, many scholars have done corresponding studies on the green development of Chinese shipbuilding industry. For example, Wen Fan studied the influencing factors of the whole process of green production in Chinese industry based on inter-provincial panel data (Fan 2019). Chunyan Jiang and Hongbo Guan calculated the green total factor productivity of Chinese shipbuilding industry using directional distance function as well as ML index, and studied the influence of annual growth rate of regional GDP, technology level, foreign export dependence, and environmental regulation on green total factor productivity of Chinese shipbuilding industry (Jiang and Guan 2021). At the same time, industrial green production can also have an impact on the cost effectiveness of enterprises. For example, Hongbin Liu focused on the cost of ship enterprises and optimized around cost in terms of planning and quality (Liu 2017). He first nalyse the trend of ship contract price based on Minitab and established Cobb–Douglas model to nalyse the impact of factors such as the level of ship production technology, the amount of labor and capital input on the efficiency of ship enterprises.

Jiangchao Wang studied the energy consumption and emission problems of shipbuilding industry in the context of increasingly severe global environmental problems (Deng 2020). He pointed out the current problems in energy and environment in shipbuilding industry and proposes corresponding energy saving and emission reduction measures to promote the transformation and upgrading of the industry. Meifei Wang investigated the relationship between the energy efficiency design index of ships and the energy consumption of shipbuilding in relation to the energy consumption and greenhouse gas emissions of shipbuilding in conjunction with the new shipbuilding energy efficiency design index formula (EEDI) (Qiu 2019).

14.2 Analysis of the Connotation and Cost Composition of Green Manufacturing in Shipbuilding Industry

14.2.1 The Connotation of Green Manufacturing

Green manufacturing is a manufacturing model that incorporates environmental awareness into the manufacturing process (Yao 2019). It is characterized by a comprehensive consideration of the environmental benefits of the manufacturing process and the economic benefits of the enterprise. The ultimate goal is to minimize the negative impact on the environment during the entire production cycle, i.e., from R&D and design to construction and maintenance, and to maximize the utilization of resources in order to achieve synergistic optimization of economic and social and environmental benefits (Li and Ni 2015). Environmental issues are a hot issue of concern around the world, and industries are responding positively to this global consensus. At present, industrial enterprises are gradually improving the traditional manufacturing model and developing relevant green manufacturing technologies to promote the transformation of industrial production to low loss, high resource efficiency, low cost and high economic efficiency.

Shipbuilding products have obvious peculiarities compared with other industries due to the large volume of products and long production cycle. It is difficult to operate the recycling of discarded shipbuilding products, which undoubtedly leads to a large amount of waste of resources and environmental pollution. Therefore, the reduction of energy consumption and pollutant emission during the design and manufacture of ships is an important way to achieve green production. Of course, the corresponding energy saving and emission reduction measures will also bring unavoidable changes in shipbuilding costs. Based on the relevant panel data of major shipbuilding provinces, municipalities, autonomous regions and municipalities directly under the central government from 2017 to 2020, the article nalysed the relationship between the main energy consumption and main pollutant emission data in the shipbuilding process and the shipbuilding cost, in order to improve the study of shipbuilding cost changes in the process of green manufacturing of ships.

14.2.2 Green Production and Cost Components in the Shipbuilding Industry

The implementation of green manufacturing in the shipbuilding industry must not only analyze its cost–benefit relationship based on production costs, but must also consider the environmental costs and the costs of managing the green environmental value chain contained in the entire life cycle of ship products from design to disposal (Yingzi and Wen 2013; Sun 2010; Suo and Wang 2008; Yan et al. 2001). The content of green environmental value chain mainly includes: (1) supply of consumables (2) product production process; (3) packaging and transportation; (4) maintenance and repair; and (5) recycling and reuse. Therefore, the cost of implementing green manufacturing in the shipbuilding industry based on environmental value chain includes: Design cost, manufacturing cost, environmental cost, marketing cost, maintenance cost, recycling and disposal cost, etc.

Design Cost. The design solution of a shipbuilding product determines the resource utilization rate and the degree of environmental impact. The ship design solution determines 80–90% of the resource consumption of a shipbuilding product throughout its life cycle.

Manufacturing Cost. The manufacturing cost of a shipbuilding product is the production cost, which includes the cost of production consumables, manufacturing process cost, ship outfitting cost, etc.

Environmental Cost. Environmental cost refers to the cost that ship enterprises need to pay to solve environmental pollution and ecological damage in the process of producing shipbuilding products. It mainly includes: the cost of controlling pollution emission, the cost of dealing with pollution emission and the cost of disposing of discharged waste.

Marketing Costs. Marketing costs include the cost of packaging, shipping and storage of ship products.

Operation and Maintenance Costs. Operation and maintenance costs include the cost of operating the shipbuilding products and the cost of repair and care.

Cost of Recycling and Disposal after Scrapping. The cost of post-waste recycling in the shipbuilding industry includes the cost of recycling ship parts and components, the cost of non-recyclable discards, and the cost of recycling parts and components to produce materials.

14.3 A Study on the Relationship Between Energy Consumption and Emissions of Shipbuilding and Cost Changes Based on Ridge Regression Model

14.3.1 Index Selection

Energy Consumption and Pollutant Emissions Indexes. Total energy consumption in the shipbuilding industry NS_i reflects the use of raw coal and crude oil by shipbuilding enterprises in the manufacturing process, excluding low calorific value fuels, biomass and solar energy, etc. Electricity consumption (billion kilowatt hours)NC_i, natural gas consumption (billion cubic meters) NT_i reflect the total amount of electricity and natural gas consumed in the manufacturing process of shipbuilding industry respectively. The emissions of ammonia nitrogen in wastewater (10,000 tons)SA_i, chemical oxygen demand in wastewater (10,000 tons)SC_i, total sulfur in wastewater (10,000 tons)SS_i, sulfur dioxide in exhaust gas (10,000 tons)SN_i, nitrogen oxides in exhaust gas (10,000 tons)SCO_i, and particulate matter in exhaust gas (10,000 tons) SK_i are the main pollutants emitted by shipbuilding enterprises in the manufacturing process. These are the main pollutants emitted by shipbuilding enterprises during the manufacturing process. Since there are no specific numerical statistics in the China Shipbuilding Industry Association and the China Shipbuilding Statistical Yearbook, this paper adopts the ratio of the output value of the shipbuilding industry SV_i to the industrial GDP V_i multiplied by the total energy consumption of the industrial industry to explain, for example.

$$NS_i = \frac{SV_i}{V_i}S_i, SN_i = \frac{SV_i}{V_i}N_i, SC_i = \frac{SV_i}{V_i}C_i, NC_i = \frac{SV_i}{V_i}C_i, ST_i = \frac{SV_i}{V_i}T_i$$

Input Cost Index. Input cost T_i is the production cost of the shipbuilding industry enterprises in the process of product manufacturing, including the cost of product design process, manufacturing process, environmental protection, marketing and maintenance, etc. This paper adopts the difference between the main business income and total profit to represent the cost value T_i, that is: $T_i = W_i - P_i$.

14.3.2 Data Pre-Processing

This paper collected cost input data as well as energy consumption and pollutant emission data of the shipbuilding industry in each province and city from 2017 to 2020. It used ammonia nitrogen emission in wastewater (10,000 tons), chemical oxygen demand in wastewater (10,000 tons), total sulfur emission in wastewater (10,000 tons), sulfur dioxide emission in exhaust gas (10,000 tons), nitrogen oxide emission in exhaust gas (10,000 tons), particulate matter emission in exhaust gas (10,000 tons), total energy consumption (million tons of standard coal), electricity consumption (billion kilowatt hours), and natural gas consumption (billion cubic meters) as independent variables, and input (revenue—profit) as dependent variables for regression analysis. Since each variable has a different magnitude and a different range of values (see Table 14.1), this paper first needs to preprocess the variables to logarithmize them.

The descriptive statistics of the logarithmic post-processed data are shown in Table 14.2 below.

According to Table 14.2, it can be seen that logarithmicizing the variables made the range of variation of each variable smaller and the difference in values between the variables narrower, which could optimize the results of the regression analysis.

14.3.3 Selection of the Regression Model

In order to study the influence of energy consumption and pollutant emissions on the change of shipbuilding cost in the shipbuilding process, this paper established a regression analysis model to investigate. Generally speaking, the problem of data covariance often occurs when regression analysis is performed on the data. The judgment criterion is that when the variance expansion coefficient of the group of data exceeds 5, it indicates that there is a covariance problem in the group of data, and if the value exceeds 10, it indicates that the covariance problem is more serious. The existence of data covariance will lead to a certain degree of bias in the data analysis results, and in serious cases, it may even lead to the opposite analysis results. As can be seen from Table 14.3, there are several groups of VIF values greater than 10 for this group of data, and the highest value has exceeded 170, which shows that the panel data to be studied has strong covariance and is not suitable for OLS regression analysis using the least squares method.

Table 14.1 Descriptive statistical analysis of variables

Name	Sample size	Min value	Max value	Average value	Standard deviation	Median
Inputs (revenue—profit) (billion yuan)	92	0.735	3379.591	434.729	723.157	127.144
Ammonia nitrogen emission in wastewater (10,000 tons)	92	0.000	0.175	0.020	0.033	0.006
Chemical oxygen demand in wastewater (10,000 tons)	92	0.000	6.484	0.377	0.866	0.131
Total sulfur emission in wastewater (10,000 tons)	92	0.000	0.067	0.005	0.010	0.001
Sulfur dioxide emission in exhaust gas (10,000 tons)	92	0.000	1.914	0.166	0.366	0.045
Nitrogen oxide emission in exhaust gas (10,000 tons)	92	0.000	4.074	0.489	0.913	0.189
Particulate matter emission in exhaust gas (10,000 tons)	92	0.000	4.568	0.344	0.815	0.079
Total energy consumption (million tons of standard coal)	92	0.244	1252.852	195.883	326.392	55.673
Electricity consumption (billion kilowatt hours)	92	0.039	215.321	29.321	50.176	8.721
Natural gas consumption (billion cubic meters)	92	0.001	9.848	1.142	2.064	0.253

The general approach to address the impact of covariance problems is to remove the data items with covariance, but this treatment will result in the target variables being unstudied. Therefore, the article used Ridge Regression analysis, which forwent the unbiased nature of least squares to find model equations that are slightly less effective but have regression coefficients that are more realistic at the expense of losing some information. In general, the robustness of Ridge Regression analysis

Table 14.2 Descriptive statistical analysis after logarithmization of variables

Name	Sample size	Min value	Max value	Average value	Standard deviation	Median
Ln_ Inputs (revenue—profit) (billion yuan)	92	−0.308	8.126	4.342	2.429	4.845
Ln_ ammonia nitrogen emission in wastewater (10,000 tons)	92	−12.429	−1.741	−5.587	2.562	−5.113
Ln_ chemical oxygen demand in wastewater (10,000 tons)	92	−8.867	1.869	−2.867	2.579	−2.036
Ln_ total sulfur emission in wastewater (10,000 tons)	92	−13.816	−2.710	−7.337	2.638	−6.609
Ln_ sulfur dioxide emission in exhaust gas (10,000 tons)	92	−12.024	0.649	−3.799	2.676	−3.110
Ln_ nitrogen oxide emission in exhaust gas (10,000 tons)	92	−8.129	1.405	−2.470	2.420	−1.665
Ln_ particulate matter emission in exhaust gas (10,000 tons)	92	−10.350	1.519	−2.987	2.496	−2.544
Ln_ total energy consumption (million tons of standard coal)	92	−1.411	7.133	3.565	2.410	4.019

tends to be stronger than linear OLS regression when there is covariance in the target data for regression analysis.

14.3.4 Principle of Ridge Regression Model and Model Building

Ridge Regression is a research algorithm used to solve the problem of covariance of independent variables in linear regression analysis (Zhang 2013; Lyu 2014; Wang 2019; Zhiyun 2020). Ridge Regression makes the regression coefficients estimable by

Table 14.3 Analysis of data covariance indexes

Model	Non-standardized coefficient		Standardized coefficient	t	Significance	Covariance statistics	
	B	Standard error	Beta			Allow	VIF
(Constant)	−2.460	12.818		−0.192	0.848		
Ammonia nitrogen emission in wastewater (10,000 tons)	9830.537	1928.266	0.444	5.098	0.000	0.023	42.875
Chemical oxygen demand in wastewater (10,000 tons)	74.642	69.446	0.089	1.075	0.286	0.026	39.142
Total sulfur emission in wastewater (10,000 tons)	−20,038.405	7083.540	−0.285	−2.829	0.006	0.017	57.419
Sulfur dioxide emission in exhaust gas (10,000 tons)	−714.674	284.317	−0.361	−2.514	0.014	0.009	116.923
Nitrogen oxide emission in exhaust gas (10,000 tons)	217.744	137.670	0.275	1.582	0.118	0.006	170.683
Particulate matter emission in exhaust gas (10,000 tons)	−3.168	70.451	−0.004	−0.045	0.964	0.028	35.630
Total energy consumption (million tons of standard coal)	0.258	0.333	0.116	0.774	0.441	0.008	127.966
Electricity consumption (billion kilowatt hours)	4.251	1.843	0.295	2.306	0.024	0.011	92.502
Natural gas consumption (billion cubic meters)	123.518	18.765	0.353	6.582	0.000	0.062	16.219

introducing a k-unit array, which displaces higher regression model rationality despite the fact that the introduction of the unit array leads to partial loss of information, and the research process is divided into 2 steps.

The first step: before the ridge regression analysis, the K value should be confirmed with the ridge plot. The K value is chosen as the minimum K value when the standardized regression coefficients of each independent variable tend to be stable; the smaller the K value, the smaller the deviation, and when the K value is 0, it is an ordinary linear OLS regression.

Step 2: For the K value, the smaller it is, the better, usually less than 1 is recommended; once the K value is determined, the K value can be actively entered to derive the ridge regression model estimate.

14.3.5 Impact of Shipbuilding Energy Consumption and Pollutant Emissions on Cost Changes

This paper used ammonia nitrogen emission from wastewater (10,000 tons), chemical oxygen demand from wastewater (10,000 tons), total sulfur emission from wastewater (10,000 tons), sulfur dioxide emission from exhaust gas (10,000 tons), nitrogen oxide emission from exhaust gas (10,000 tons), particulate matter emission from exhaust gas (10,000 tons), total energy consumption (10,000 tons of standard coal), electricity consumption (billion kilowatt hours), and natural gas consumption (billion cubic meters) as independent variables, and input cost (revenue-profit) as dependent variables for Ridge Regression analysis. Firstly, no K values were entered to obtain ridge trace plots for judging the best K values, and the ridge trace plots are shown below.

According to the above Fig. 14.1, it can be seen that when Ln _total output value, n _Total sulfur emissions from wastewater (million tons), Ln _Sulfur dioxide emissions from waste gas (million tons), Ln _Nitrogen oxide emissions from waste gas (million tons), Ln _Particulate matter emissions from waste gas (million tons), Ln _ Total energy consumption (million tons of standard coal), Ln _Electricity consumption (billion kWh), Ln _Natural gas consumption (billion cubic meters) are seen as independent variables, and the Ridge trace plot obtained after Ridge Regression analysis with Ln _inputs (revenue-profit) (billion yuan) is used as dependent variables, it can be seen that the standardized regression coefficients of independent variables tend to be stable at this time when the value of K was 0.99, thus the optimal K value was taken as 0.99.

Taking the value of K as 0.990, it can be seen from the table above that the model R^2 value of 0.931 means that the emissions of ammonia nitrogen from Ln _wastewater (million tons) (see Table 14.4), chemical oxygen demand from Ln _ wastewater (million tons), total sulfur from Ln _wastewater (million tons), sulfur dioxide from Ln _exhaust gas (million tons), nitrogen oxides from Ln _exhaust gas (million tons), particulate matter from Ln _exhaust gas emissions (million tons), Ln

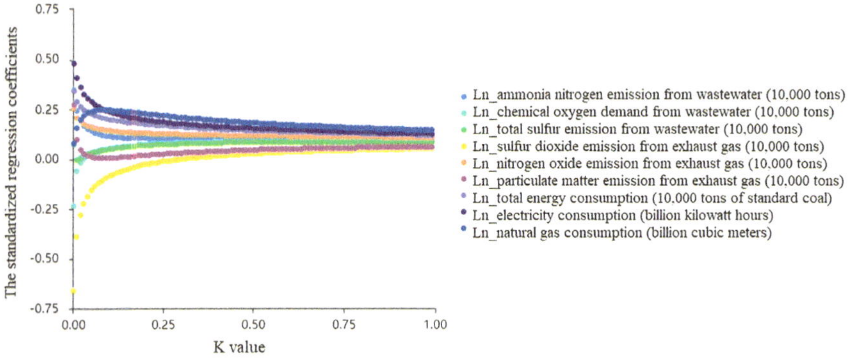

Fig. 14.1 Ridge trace map

_Total energy consumption (million tons of standard coal), Ln _Electricity consumption (billion kWh), Ln _Natural gas consumption (billion cubic meters) can explain 93.15% of the variation of Ln _Inputs (revenue - profit) (billion yuan).

The model passed the F-test ($F = 123.895$, $p = 0.000 < 0.05$), which means that at least one of the nine shipbuilding energy consumption and emission data selected in the paper has an impact on the cost of shipbuilding. The model is valid and a table of coefficients of the ridge regression analysis is obtained after conducting a ridge regression analysis on it (see Table 14.5).

According to Table 14.6, the Ridge Regression model equation is:

$$\ln T_i = 5.912 + 0.090 * \ln SA_i + 0.083 * \ln SC_i + 0.081 * \ln SS_i + 0.049 * \ln SN_i \\ + 0.106 * \ln SCO_i + 0.059 * \ln SK_i + 0.123 * \ln NS_i + 0.126 * NC_i + 0.151 * NT_i$$

Based on the significant levels of each variable in the above table, it is clear that the value of the regression coefficient of ammonia nitrogen emissions in Ln_ wastewater after logarization of the data was 0.090 ($t = 16.032$, $p = 0.000 < 0.01$), implying that ammonia nitrogen emissions in Ln_ wastewater can have a significant positive relationship on Ln_ input costs. The regression coefficient value for COD in Ln_

Table 14.4 Model summary for ridge regression analysis

R^2	*Adjust* R^2	Model error RMSE
0.931	0.924	0.632

Table 14.5 Results of variance analysis for ridge regression analysis

	Quadratic sum	df	Mean square	F	p value
Regression	500.032	9	55.559	123.895	0
Residual	36.772	82	0.448		
Total	536.804	91			

Table 14.6 Table of coefficients of ridge regression analysis

	Non-standardized coefficient		Standardized coefficient	*t*	*p*
	B	Standard error	*Beta*		
Constant	5.912	0.118	-	50.186	0.000**
Ln_ ammonia nitrogen emission in wastewater (10,000 tons)	0.090	0.006	0.095	16.032	0.000**
Ln_ chemical oxygen demand in wastewater (10,000 tons)	0.083	0.006	0.088	13.494	0.000**
Ln_ total sulfur emission in wastewater (10,000 tons)	0.081	0.007	0.088	12.179	0.000**
Ln_ sulfur dioxide emission in exhaust gas (10,000 tons)	0.049	0.007	0.054	7.179	0.000**
Ln_ nitrogen oxide emission in exhaust gas (10,000 tons)	0.106	0.005	0.106	23.439	0.000**
Ln_ particulate matter emission in exhaust gas (10,000 tons)	0.059	0.008	0.061	7.481	0.000**
Ln_ total energy consumption (million tons of standard coal)	0.123	0.005	0.123	25.586	0.000**
Ln_ electricity consumption (billion kilowatt hours)	0.126	0.005	0.129	25.030	0.000**
Ln_ natural gas consumption (billion cubic meters)	0.151	0.009	0.146	16.060	0.000**

wastewater was 0.083 ($t = 13.494$, $p = 0.000 < 0.01$), implying that COD in Ln_ wastewater would have a significant positive relationship on Ln_ input cost. The regression coefficient value of 0.081 ($t = 12.179$, $p = 0.000 < 0.01$) for total sulfur emissions in Ln_ wastewater implies that total sulfur emissions in Ln_ wastewater can have a significant positive relationship on Ln_ input costs. The regression coefficient value of 0.049 ($t = 7.179$, $p = 0.000 < 0.01$) for sulfur dioxide emissions in Ln _ exhaust gas implies that sulfur dioxide emissions in Ln _exhaust gas will have a significant positive relationship on Ln _input costs. The regression coefficient value of 0.106 ($t = 23.439$, $p = 0.000 < 0.01$) for nitrogen oxide emissions in Ln _exhaust gas implies a significant positive relationship between nitrogen oxide emissions in Ln _exhaust gas and Ln _input costs. The regression coefficient value of 0.059 ($t = 7.481$, $p = 0.000 < 0.01$) for particulate matter emissions in Ln _exhaust gas implies a significant positive relationship between particulate matter emissions in Ln _exhaust gas and Ln _input costs. The value of regression coefficient for total Ln _energy consumption was 0.123 ($t = 25.586$, $p = 0.000 < 0.01$), implying that total Ln _energy consumption will have a significant positive relationship on Ln _ input costs. The value of the regression coefficient of Ln _power consumption was 0.126 ($t = 25.030$, $p = 0.000 < 0.01$), implying that Ln _power consumption will have a significant positive relationship on Ln _input costs. The value of regression coefficient for Ln _gas consumption was 0.151 ($t = 16.060$, $p = 0.000 < 0.01$), implying that Ln _gas consumption will have a significant positive relationship on Ln _input cost.

To sum up the analysis, it can be seen that the selected energy consumption and emission indexes in the shipbuilding process have a positive relationship with the shipbuilding cost, and they are, in order of influence, natural gas consumption, electricity consumption, energy consumption, nitrogen oxide emission in waste gas, ammonia nitrogen emission in waste water, chemical oxygen demand emission, sulfur compound emission in waste water, particulate matter emission in waste gas, and sulfur dioxide emission in waste gas.

14.4 Conclusion and Countermeasures

Based on the panel data of the shipbuilding industry in major provinces, municipalities and autonomous regions in recent years, this study integrated the indexes of energy consumption and pollutant emission, firstly pre-processed the data logarithmically, and then used a Ridge Regression model to study the dynamic change relationship between shipbuilding cost and energy consumption and pollutant emission. The article gave the corresponding influence coefficients for the degree of influence of different factors on the change of shipbuilding cost. In the process of transformation to green manufacturing, shipbuilding enterprises can analyze the influence of the application of relevant environmental protection measures on the change of shipbuilding cost accordingly, so as to better grasp the dynamic balance between green development and economic benefits.

Meanwhile, the impact of different types of energy consumption on environmental and economic benefits in shipbuilding process is different. The impact coefficient can fuzzy quantify the degree of impact. Shipbuilding companies can adjust energy consumption according to demand to reduce production costs and maintain environmental benefits, or develop green energy instead of traditional energy to reduce pollutant emissions.

Comprehensive research in this paper shows that China's shipbuilding industry, under the current background of green economy and green industry, should face the market demand and the practical needs of the people and keep vigorously developing green shipbuilding. At the same time, shipbuilding enterprises should also actively explore the economic issues of shipbuilding and environmental issues such as energy saving and emission reduction in the context of green manufacturing. Only by achieving both green manufacturing and enterprise development can shipbuilding enterprises better realize the modernization of shipbuilding enterprises and achieve the sustainable development of green economy.

14.5 Discussion and Deficiency

Due to the problem of data availability, the data selected in the article are the panel data of the major provinces in China in the last four years related to ship production and manufacturing, and the subsequent study can appropriately increase the amount of data, including adding years, provinces and pollutant emission types, in order to further improve the accuracy of data analysis. The article adopted the data analysis method of ridge regression, which inevitably led to a certain degree of result data bias due to the introduction of unit matrix. The subsequent research process can appropriately adjust the ridge parameter (K value) to verify the reasonableness of the conclusion, and other more scientific regression analysis methods can also be studied to do the conclusion optimization.

References

Cuicui Sun (2010) Cost-effectiveness study of green manufacturing in the steel industry. Shandong Univ Financ Econ

Dingsan Qiu (2019) Study on industrial green production efficiency and its influencing factors in the Yangtze River Economic Belt. Anhui Univ Sci Technol

Haixin Lyu (2014) Research on the cost management of ship manufacturing enterprises based on supply chain. Jiangsu Univ Sci Technol

Hongbin Liu (2017) Shipbuilding project management and cost-benefit analysis. South China Univ Technol

Jiang C, Guan H (2021) Chinese shipbuilding green total factor productivity and its determinants: Retesting the "Porter Hypothesis" from the perspective of ML productivity. Marine Economy 11(02):62–71

Jing Li, Dongxue Ni (2015) China's industrial green production and governance efficiency research—Based on two-stage SBM network model and global Malmquist method industrial economics research, 03, pp 42–53

Li Zhang (2013) Cost control study of ship structure manufacturing. Chin Foreign Entrep, 04, pp 02

Pinping Wang (2019) Study on the whole life cycle cost management of marine engine sales project of enterprise A. Univ Electron Sci Technol China

Suo G, Wang Y (2008) Study on the strengthen policy of China's iron and steel industry based on endogenous innovation and green manufacture. Sci Res Manag 04:30–33

Wang H (2021) Liaoning shipbuilding industry high-quality development to promote intelligent manufacturing, intelligent shipping and green maintenance. China Ship Repair 34(06):03

Wenmin Deng (2020) Dynamic evaluation of provincial green production efficiency based on Network Global Malmquist-Luenberger Index. Nanjing Univ Posts Telecommun

Wen Fan (2019) Mechanisms and empirical studies of government subsidies driving industrial green growth. Hunan University

Yan H, Ren P, Li X, Ye R (2001) Cost-effectiveness of green production: SMP case study. Reform Econ Syst 04:89–91

Yue Yao (2019) A study of China's industrial green production performance and the impact of energy saving and emission reduction targets on it. Hefei Univ Technol

Yingzi Wu, Yuechun Wen (2013) Green productivity, deduction performance, and abatement cost for the industry in China. Sci Res Manag, 34 (02), pp 105–111+151

Zhiyun Lu (2020) Project cost management analysis and management thinking of shipbuilding enterprises. Chin & Foreign Entrep, 07, pp 02

Chapter 15
Analysis and Research on Basic Structure of the PID Steam Pressure System

Xiaolin Gao and Jing Lu

Abstract In order to improve the working efficiency of the steam system, it is necessary to optimize the operating parameters of the steam system in an all-round way. At the same time, it can also provide theoretical basis and technical support for energy conservation and environmental protection. This will be what enterprises must face. Due to the characteristics of multivariable, nonlinear, large delay, etc., the regulation of industrial boilers is also complex. For the control process, it is necessary to use PID for regulation, which can effectively reduce the overshoot, shorten the regulation time, reduce the fluctuation of required control parameters, and reduce the impact on the control performance, which will directly affect the stability of the system.

Keywords Steam pressure · PID · Automatic control

15.1 Introduction

Industrial boiler system is a very important system in industrial production. The most important parameter for the operation safety of industrial boilers is the steam pressure. Because industrial boilers are closely related to people's daily life, it is very important to control the stable operation of steam pressure. In the steam control system, if the pressure bearing parts can not bear too high air pressure, explosion and other dangerous accidents may occur, which will bring extremely serious consequences to the operators and equipment; If the steam pressure in the steam control system is lower than the minimum standard, the coal consumption and steam consumption will increase sharply, which will seriously interfere with power generation and heating and increase its cost; The low-frequency oscillation of the power grid is directly related to the stability of the steam pressure. A little change in the steam system will lead to a chain reaction. In serious cases, it will lead to the overall paralysis

X. Gao (✉) · J. Lu
Heilongjiang University of Technology, Heilongjiang, China
e-mail: 252659020@qq.com

J. Baeyens et al. (eds.), *Proceedings of 2022 4th International Conference on Environment Sciences and Renewable Energy*, Environmental Science and Engineering, https://doi.org/10.1007/978-981-19-9440-1_15

of the system, resulting in power failure in the region. In order to ensure the safety of production, it is necessary to maintain the boiler steam pressure within a certain range. For thermal power plants, chemical plants, boiler plants and steam turbine plants with relatively strict steam pressure control, if the pressure is too low, it will not meet the requirements, and if the pressure is too high, it is easy to damage the equipment. Therefore, the most important thing is to effectively regulate and control the steam pressure, which is extremely important for production and life (Zhengjianping 2006).

The power required by the steam turbine is mainly provided by the steam pressure difference. Therefore, it is necessary to improve the utilization of the pressure system and reduce the pressure to be borne by the pipeline. Negative pressure phenomenon will appear in the vacuum circuit, which is particularly important for controlling the steam pressure of the main extraction valve. The steam pressure shall be continuously monitored in real time. During the operation of the equipment, the excessive steam pressure will lead to the increase of mechanical stress and damage to the steam pipeline and boiler, which will reduce the safety performance and economy of the steam turbine and boiler. On the contrary, if the steam pressure is too low, the expansion work of steam will be reduced, resulting in excessive steam loss. Generally, the lower limit of steam pressure is 6% of the rated value. If it is lower than the lower limit, the steam consumption will increase by 1% or more. At this time, the equipment will work under the load reduction state, that is, abnormal operation will lead to equipment damage. In view of this, this paper must monitor the actual value of steam pressure in real time in order to control and maintain its stability (Wang et al. 2005).

In order to ensure the safe operation of real gas pressure control and the controllable automation level, it is necessary and significant to adopt more advanced algorithm design, real-time monitoring and control of automatic equipment.

15.2 Analysis and Research on the Basic Structure of Steam Pressure Controller

15.2.1 Digital PID Controller

Digital PID control is a widely used and advanced control method in the current industrial control industry. Its control characteristics can ensure the accuracy, stability and rapidity of the control system. PID controller can also be combined with intelligent control such as neural network control to improve control accuracy and improve control methods. It is the most extensive control method in the control system.

Advantages of PID Controller. In the industrial control, PID is the most common control method. P means proportion, I means integral, and D means differential. PID controller is the most widely used and advanced in the field of closed-loop control. The reason why PID controller is so widely used is that it has many advantages:

There is No Need to Calculate the Mathematical Model of the Controlled Object. Theoretically, the automatic control system needs to get the linear steady mathematical model of the controlled object, and the accurate mathematical model can get the accurate control effect. In reality, most of the control objects are nonlinear and time-varying. Even if the mathematical method is used to convert them into linear models, there is still a large error, and only approximate control can be carried out. Therefore, in the actual application control system, the control method of automatic control principle is not effective. For most control systems, PID control method can meet the control requirements of this paper.

Simple Structure and Easy Implementation. The structure of the PID controller is relatively simple, the program design is relatively easy, no tedious mathematical calculation is required, the parameters in the model are clearly defined, only the parameters need to be adjusted to change the control effect, and complex control systems such as multi-loop and cascade control can also be realized.

Strong Control Flexibility and Adaptability. Because the three control modes can be combined and converted arbitrarily, they can be adjusted according to the characteristics of the controlled object or different control requirements, such as PD, PI, integral separation PID and variable speed integral PID. As intelligent control is gradually applied to the control system, PID control method can also be combined with modern control methods such as neural network control and fuzzy control. PID control parameters can also be self-tuning. Therefore, in the industrial control industry, PID control method has a broad prospect and application range.

Easy to Use. At present, many controllers have their own internal integrated PID controllers. The products will have their own PID closed-loop control modules, PID control system function modules, PID control instructions, etc., which are convenient for users. According to different control effects, it is only necessary to reset the parameters, and many products can realize parameter self-tuning (Water and steam 2008).

15.2.2 Digitalization of PID Controller

Expression of PID Controller in Continuous System. The transfer function of PID controller is

$$\frac{MV(s)}{EV(s)} = K_p\left(1 + \frac{1}{T_{I^s}} + T_{D^s}\right) \tag{15.1}$$

The output expression of the analog PID controller is

$$mv(s) = K_P\left[ev(t) + \frac{1}{T_I}\int ev(t)dt + T_D\frac{dev(t)}{dt}\right] + M \tag{15.2}$$

In the expression, it is mainly divided into three parts: proportional, integral and differential. The input of the controller is the control error signal $ev(t) = sp(t) - pv(t)$; *sp(t)*is the initial value; *pv(t)*is the actual value; *mv(t)* is the output of the controller; K_p is the proportional coefficient; T_i is the integral time constant, T_D is the differential time constant; *M* is the initial value of the integration link.

In the expression (15.2), the first three items in brackets are proportional, integral and differential parts, which are respectively related to the error *ev(t)*, The integral of the error is proportional to the differential of the error (Wang Shitong and hybrid learning algorithm based on CM algorithm that can converge globally with probability. 2008). If the system only needs proportional, proportional integral, proportional differential and other controls, one or two of them shall be taken. If the system requires high stability and dynamic characteristics, the proportional integral control method is used; If the controlled object has large hysteresis, it is necessary to select PID control.

Approximate Calculation of Integral Part. Set the sampling period for executing PID control is *Ts*,the *n* times PID operation time is *Tsn*,because *Ts* is constant when the PID program is running, the input of PID controller *ev(Tsn)* can abbreviated as *ev(n)* when $t = Tsn$, the output value *mv(Tsn)* can abbreviated as *mv(n).*

Approximate Calculation of Differential Part. In expression (15.3),the differential expansion into difference form

$$\frac{dev(t)}{dt} \approx \frac{\Delta ev(t)}{\Delta t} = \frac{ev(n) - ev(n-1)}{T_s} \tag{15.3}$$

In the expression, ev(n-1)is the error value in the n-1 time sampling, Substituting the differential expansion of integral and differential into the expression (15.2),The output of the control system at the nth sampling is shown in the expression (15.4).

$$mv(n) = K_p\left\{ev(n) + \frac{T_s}{T_I}\sum_{j=1}^{n} ev(j) + \frac{T_D}{T_s}[ev(n) - ev(n-1)]\right\} + M \tag{15.4}$$

In the SFB41 "CONT_C" (Continuous controller),KP,TI,TD and M corresponding to input parameters respectively GAIN,TI,TD and integral initial value I_ITLVAL.

Taking the electric control valve as an example, the control system uses the position sensor to monitor the valve opening in real time, and uses the closed-loop control system to compare the valve opening with the output of the PID controller for control. In the expression (15.4), the valve opening corresponds to the output value of the PID controller, which is called the PID position algorithm.

Incomplete Differential PID. When the control system has high requirements for dynamic characteristics, the differential control method is often used to improve, but high-frequency interference is often introduced. Therefore, the first-order inertial filter is generally connected to reduce the overshoot of the output value. The transfer function of the incomplete differential PID controller given below is shown in expression (15.5):

$$\frac{MV(s)}{EV(s)} = K_p\left(1 + \frac{1}{T_I S} + \frac{T_D S}{T_f S + 1}\right) \tag{15.5}$$

Application of Dead Time Characteristic in PID Control. In the control system, for the actuator, continuous action is required, which will produce large mechanical wear. For the control system with low control accuracy requirements, the PID control method with dead zone can be used. When the input value of the nonlinear link with dead zone is between the set values, the system output is zero. At this time, the system only adopts the integral part, cancels the proportional and differential links, and the control system becomes an open-loop control system. When the input value exceeds the set value of the dead zone, the system changes back to the PID controller, which is the closed-loop control.

15.2.3 Main Steam Pressure Controller

The main steam pressure control system can effectively control the main steam pressure in the initial mode. Another function is to ensure that the actual value of main steam pressure is not higher than the preset pressure value in the pressure limit mode, which is common in boiler control.

The main steam pressure is controlled by the controller with PID control function, and the air inlet module is controlled by the central control module to control the steam flow of the whole steam turbine.

The input of PID controller is the control deviation of main steam pressure, and the control deviation is the deviation between the set value of main steam pressure and the actual value detected in real time. Once the control deviation occurs, the output of the controller is to the central min function module of the steam inlet set value module, and the set values of all control valves are set according to the central processing module. When the steam flow decreases, all valves are required to be closed in order to prevent the pressure difference in the high-pressure part of the turbine from causing danger.

When the main steam pressure controller is disabled, if the confusion between the main steam pressure controller and the current working controller is to be realized, the output value of the main steam pressure controller is the deviation increment of the central module and the steam pressure control system.

The positive and negative control deviation are related to the main steam pressure. When the main steam pressure is lower than a certain allowable value, the deviation is negative. In order to control the pressure drop, close the control valve properly. The steam inlet setpoint module activates the pressure controller and outputs the signal to the OM system. The main steam pressure controller performs relevant actions.

When the system receives the closing signal sent by the clutch, the lower output limit of the steam pressure controller is converted from the output limit signal 1 of the lower main steam pressure controller to the output limit signal 2 of the main steam pressure controller.

About HP pressure controller, two operation modes are introduced in detail:

When the steam pressure controller is in the limited pressure working mode, when the steam pressure is lower than a certain set value, the steam controller acts. In this mode, the pressure controller closes the control valve appropriately to meet the set value. HP bypass pressure controller plays a major role.

When the initial pressure is in the working mode of the initial pressure, the t pressure controller acts to reset the set value of the unit pressure and set the bypass HP pressure in front of the turbine. The specific process is to add the closing parameter " + 7 bar" in front of the set value of the unit to make HP in the closing state.

In the initial state, the pressure control valve will increase with the increase of steam flow. When the steam flow does not increase to 100%, all the control valves will be opened, and the control valve can not meet the needs of steam flow. In this way, the boiler load will continue to increase, and the bypass pressure HP will also increase.

When the steam flow load of all open valves reaches 100% load, t control valves shall remain open. At this time, the turbine with sliding pressure control will work.

Because the control section with nozzle is canceled, the system has no throttling loss under high load working conditions, so the control system with sliding pressure control is designed: take the power plant with two boilers and one steam turbine as Beijing, one of the boilers operates at full load, and the bypass pressure HP reaches 100%. At this time, the two boilers must be maintained at the maximum flow, and the main engine control unit sets the bypass pressure HP value, For the controller inside the boiler, it is not necessary to change any value.

Analysis of PID in Steam Pressure Control System. The PID control system can be expressed and analyzed by transfer function, state equation and differential equation. The transfer function is modeled to analyze the PID control system. As shown in Fig. 15.1, it is the transfer function of the closed-loop control system with negative feedback. It is the input signal of the system, the input signal of the controller, the transfer function of the controller, the transfer function of the control system, and the output of the control system.

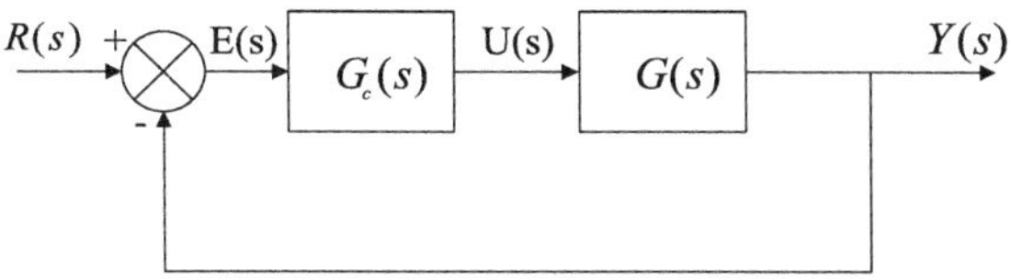

Fig. 15.1 Negative feedback closed loop control system

Due to the different product requirements and complex control objects in industrial applications, PID controller also has many structures, and corresponding adjustment methods for KP, Ki and KD parameters in PID have corresponding setting methods. According to different control objects, a suitable design scheme is designed, and a relatively perfect modeling method suitable for the control quantity in this paper is determined by the basic control structure of PID. Expression of PID controller:

$$u(t) = K_p e(t) + K_i \int_0^t e(t)dt + K_d \frac{de(t)}{dt} \tag{15.6}$$

In the expression, $u(t)$ is the output value, $e(t)$ is the input value, K_p, K_i, K_d is the adjustment parameters for the controller. The transfer function of the integer order PID controller is:

$$G_c(s) = \frac{u(s)}{e(s)} = K_p + \frac{K_i}{s} + K_d s \tag{15.7}$$

The discrete PID controller is:

$$u(k) = k_p e(k) + k_i \sum_{j=0}^{k} e(j)T + k_d \frac{e(k) - e(k-1)}{T} \tag{15.8}$$

where k is the sampling sequence number and T is the sampling period. According to the actual needs of the project and the requirements of the control system, the PID controller is divided into incremental PID, separated integral PID, PI, PD and PID. PID controllers with a wide variety of filters are often used.

15.2.4 Influence of Control Parameters on Steam System Performance

Influence of Value K_p on Steam Pressure Control System.

As shown in Fig. 15.2a, the step response of the system is affected by K_P. The larger the value of K_P, the more times the system oscillates, the greater the overshoot of the system, the slower the adjustment speed of the system, and on the contrary, the faster the response of the system. As shown in Fig. 15.2b, for the system that K_i and K_d are confirmed, the value K_P changed alone has no obvious impact on the system.

Influence of Value K_i on Steam Pressure Control System. As shown in Fig. 15.3a, the step response of the value K_i to the system. When the value K_i is very small, the steady-state error of the system will increase. If the value K_i exceeds a certain range, the stability of the system will be reduced. Therefore, it should be kept within a certain range. After setting the system, the sum value K_d and K_P remains unchanged and the value becomes smaller, which has little impact on the system. *Similarly*, the *change* of the integer K_i controller to the system and the change of the control parameters K_i of the fractional order $PI^\lambda D^\mu$ controller to the system performance are roughly the same (Research Report on optimization scheme of low pressure steam system Shanghai 2005).

Influence of Value K_d on Steam Pressure Control System. As shown in Fig. 15.4a, the impact of value K_d change on the system. When K_d beyond a certain range, the stability of the system is reduced. If K_d is too small, the overshoot of the system will become larger and the adjustment time will become longer. After K_P and K_i is set, the value K_d changes little, it has little impact on the system. At the same time, the changes of K_d in the integer order PID controller and the fractional order $PI^\lambda D^\mu$ controller have little effect on the system performance (Matyas 2005; Petras et al. 2000).

Through the analysis of the above three parameters, it can be seen that the parameters have a great impact on the same performance. Finally, we choose $Kp = 1.3$, $Ki = 2.25$, $Kd = 3.85$. This set of parameters is the optimal effect of the system. (Benjamin 2004; Qingshan and caoguangyi, wangzhenbin, 2004; Steam and pressure control system of marine boiler based on BP neural network PID. 2018).

15.3 Conclusion

This paper mainly analyzes, designs and studies the basic structure of PID steam pressure system. This paper expounds the basic structure of the steam pressure controller, expounds the PID controller selected in the steam pressure control system, analyzes its advantages and digital process, and provides a control algorithm for the steam pressure control system. It also shows that the design has certain social significance and practical value. This is not only the embodiment of the progress of the times, but also in line with scientific outlook on development of our country.

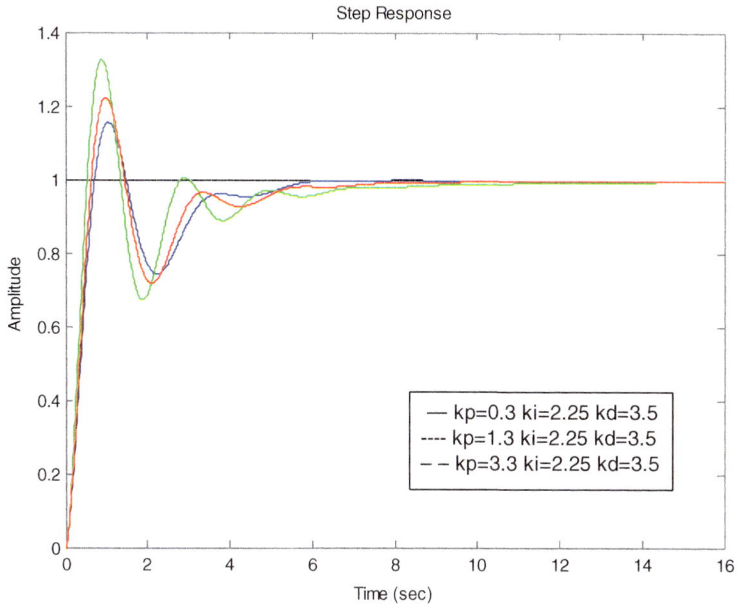

(a)

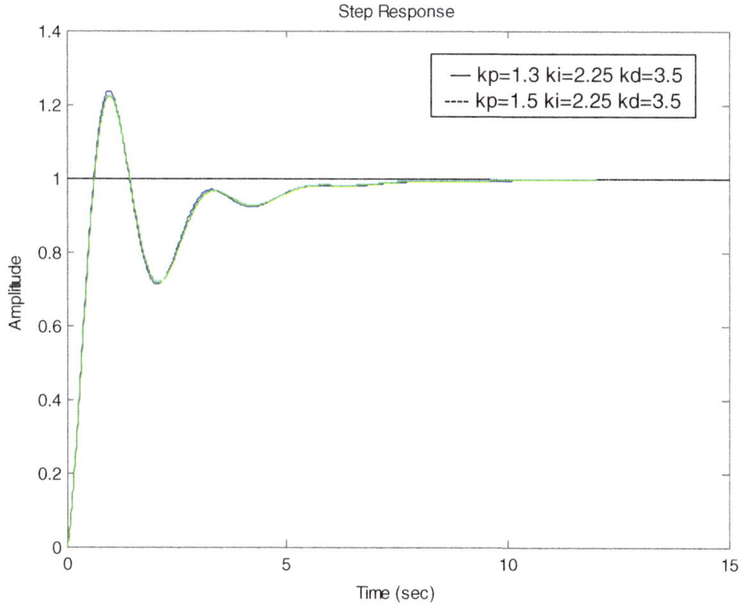

(b)

Fig. 15.2 Step response of the system when control parameters when K_P changed

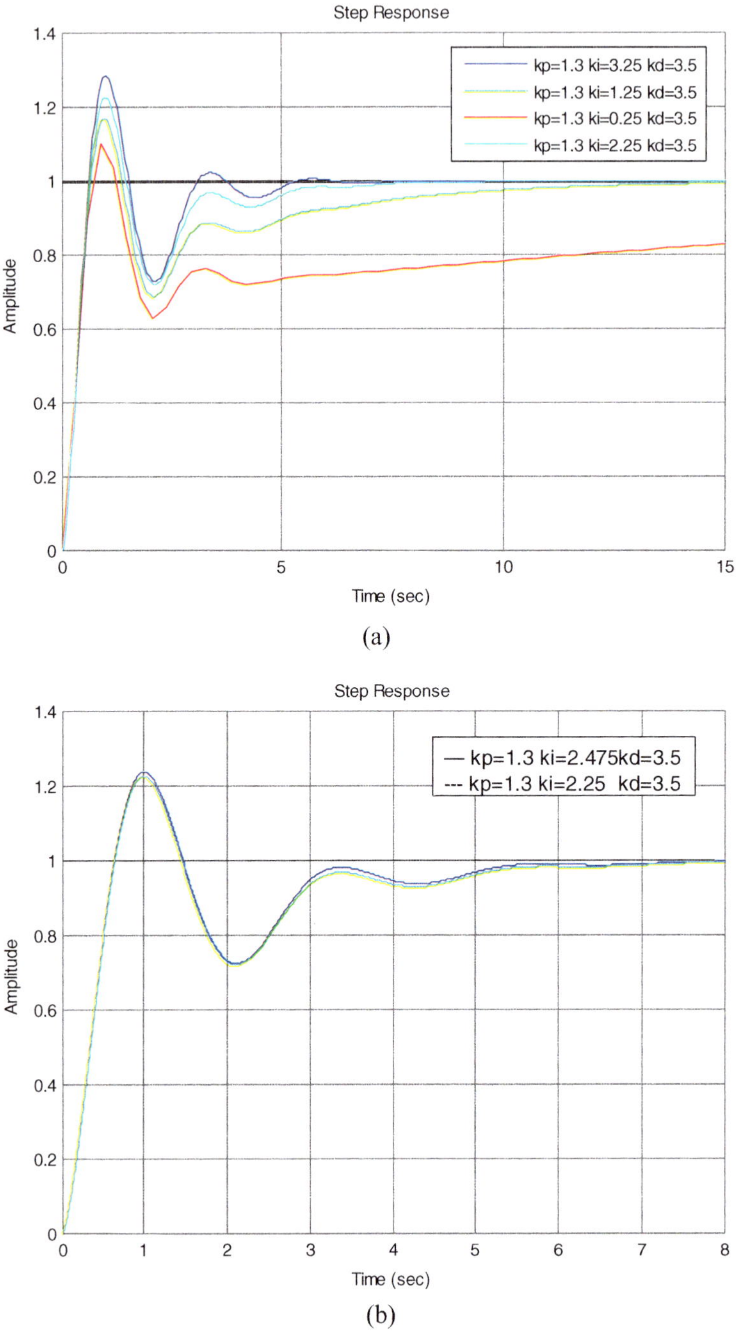

Fig. 15.3 Step response of the system when control parameters K_i change

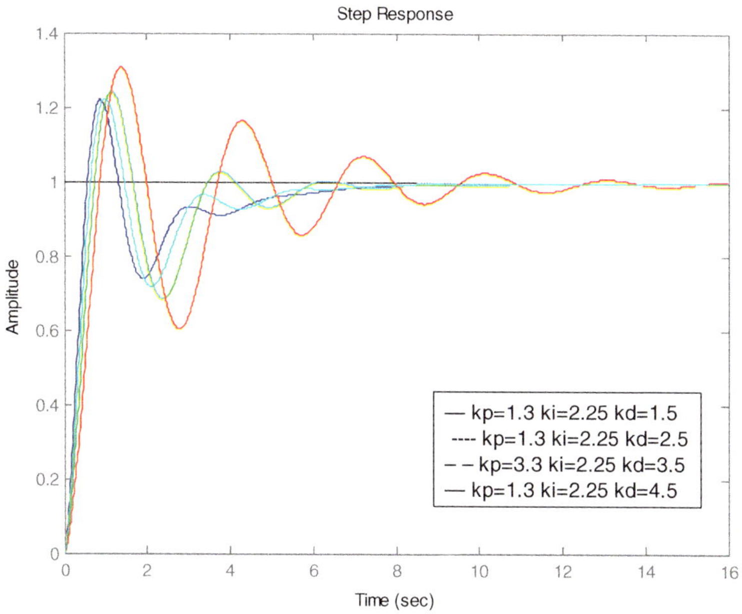

(a)

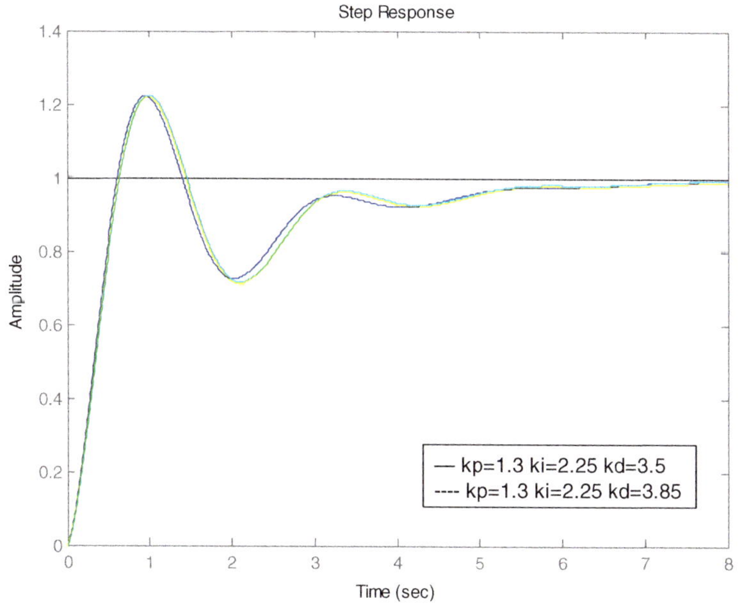

(b)

Fig. 15.4 Unit step response of the system when control parameters K_d change

References

Benjamin C Kuo (2004) Farid Golnaraghi. Automatic control system Wangxiaofan, Li Xiang Eighth edition Higher Education Press, 7: 397-518

Lujiaqi Steam pressure control system of marine boiler based on BP neural network PID. Chinese and foreign ship science and technology, 2018 (13): 95–96

Matyas J (2005) Random optimization. Autom Remote Control 28:244–251

Petras I, Dorcak L, Oleary P, et al (2000) The modelling and analysis of fractional-order control systems in frequency domain. In: ICCC'2000 Conference, May 23–26, High Tatras, pp 261–264

Research Report on optimization scheme of low pressure steam system Shanghai University of technology, 2005

Wang FS, Juan WS, Chan CT (2005) optimal tuning of PID controllers for single and cascade control loops. Chem Eng Union 132:15–34

Wang Shitong A hybrid learning algorithm based on CM algorithm that can converge globally with probability. J Softw 2008, 6(9)

Water and steam piping materials and test pressure standards Cb250–77 (2008)

Zeng Qingshan, Caoguangyi, Wangzhenbin, etc (2004) Simulation of fractional order controller journal of system simulation, issue III:456–473

Zhengjianping (2006) Huang Weijian Strategies for improving main steam pressure stability of thermal power units. J Shanghai Electr Power Inst, 22 (2): 117–120

Chapter 16
Hydrogen Production by Catalytic Conversion of Ammonia

Yimin Deng, Raf Dewil, Tom Schroeyens, Shuo Li, and Jan Baeyens

Abstract The production of hydrogen using "green" ammonia was evaluated in this work. The endothermic decomposition of ammonia into H_2 and N_2 using non-noble metal catalysts showed very promising results. "Green" ammonia sources were investigated, such as ammonia stripping from manure and digestate. H_2 will play in the future of transportation, in the energy sector and in the industry to make the processes more carbon neutral. Possible catalysts were investigated, leading to the best suitable catalyst being Fe_3O_4 supported on γ-Al_2O_3 with a 20/80 weight % ratio. This catalyst was found to be the best in price-quality due to the low cost of iron and the high activity. The appropriate reactor setup was also further assessed by an Aspen Plus simulation using a fluidized bed reactor.

Keywords Hydrogen · Ammonia · Catalysis · Conversion · Efficiency

16.1 Introduction

16.1.1 The Demand and Potential of Hydrogen

When dealing with the valorization of hydrogen, two main issues need to be considered, i.e. the growing market for H_2; and the "green" feedstock that should be used to produce H_2. The demand for H_2 continues to increase. It is almost entirely supplied from fossil fuels. The market value was 755 M$/year in 2020. It is expected to

Y. Deng · R. Dewil · T. Schroeyens · J. Baeyens (✉)
KU Leuven, Department of Chemical Engineering, Environmental and Process Technology Lab, 2860 Sint-Katelijne-Waver, Belgium
e-mail: baeyens.j@buct.edu.cn

S. Li · J. Baeyens
Beijing Advanced Innovation Center for Soft Matter Science and Engineering, Beijing University of Chemical Technology, Beijing 100029, China

J. Baeyens et al. (eds.), *Proceedings of 2022 4th International Conference on Environment Sciences and Renewable Energy*, Environmental Science and Engineering, https://doi.org/10.1007/978-981-19-9440-1_16

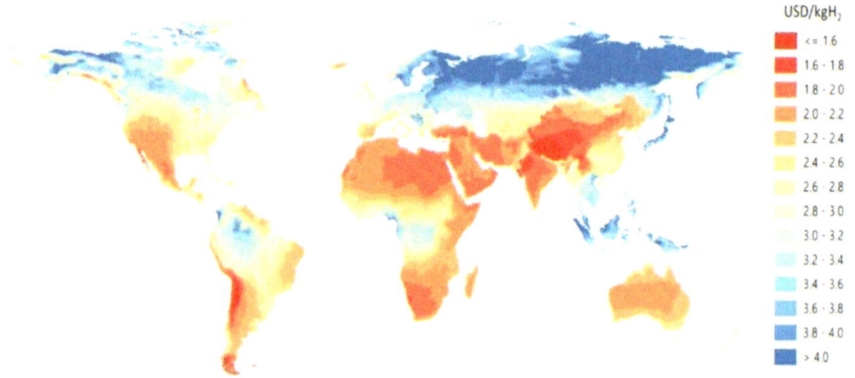

Fig. 16.1 Long term H_2 costs from hybrid solar PV and onshore wind systems (2019)

increase to 1423 M$ by 2026 (2026). Being fossil fuel-based, the production of H_2 is responsible for CO_2 emissions of around 830 million tons per year.

Major industrial users will be the steel (Li et al. 2021a), and the limestone/cement (2022) industries, responsible for nearly 15% of the anthropogenic CO_2 emissions. But the energy and transportation sectors will largely benefit from a higher availability of H_2. Since costs for solar PV and wind generation are decreasing, building electrolysers at locations of renewable electricity resources can become a low-cost supply option for hydrogen, even accounting for the transmission and distribution costs of H_2 from (often remote) renewables locations to the end-users. Figure 16.1 illustrates the long term H_2 costs from hybrid solar PV and onshore wind systems.

For transportation, the development of fuel cell technology merits a special attention. H_2 and fuel cells are now regarded as key transportation solutions beyond 2030. Although outside the scope of the present research, literature has cited a work-efficiency of fuel cells beyond 60%, thus far above the 30–35% achieved in a petrol-fueled vehicle. If using H_2 as fuel cell feedstock, only H_2O will be produced, thus avoiding the emissions of CO, CO_2 and NO_x from the traditional car engines.

16.1.2 Different Hydrogen Production Methods

Hydrogen Derived from Fossil Fuels. Currently 96% of all hydrogen is derived from fossil fuels, with the major feedstock being methane (49%), liquid hydrocarbons (29%) and coal (18%). Methane can be used for hydrogen production in 4 different pathways: Steam reforming of methane, its partial oxidation, its auto-thermal reforming, or its catalytic decomposition. The steam reforming of methane is a very mature and cheap process but it yields large amounts of CO_2. Partial oxidation combusts methane to CO and H_2, the remaining CO can be converted by the water–gas-shift reaction. Auto-thermal reforming is a steam reforming with an O_2

supply. The catalytic decomposition of CH_4 yields H_2 and C, and no CO or CO_2 is produced. External energy is needed because of the endothermic reaction. It is less polluting and the by-product, carbon black, can also be sold (Deng et al. 2022).

Water Electrolysis. Water can be split into H_2 and O_2 using electricity by 3 possible technologies: alkaline electrolysis (AEL), polymer electrolyte membrane electrolysis (PEM) and solid oxide electrolysis (SOE). AEL is already in commercial use and is mature and stable. The electrolyte is however corrosive and the process has slow dynamics. PEM produces high purity H_2, has a compact design and fast start-up. Disadvantages are the use f noble metals due to the acidity and expensive polymeric membranes. The technique is very near commercial use. High-temperature SOE is still on laboratory scale with expected low capital costs, enhanced thermodynamics and kinetics, and lower energy demands. The major SOE problems are the microstructural changes in the electrodes and the safety issues due to improper sealing.

Biomass Derived Hydrogen. Biomass, wood, agricultural residues, food waste, energy crops, and sewage sludge can be used for the production of hydrogen. Applied methods are mostly fermentation, biomass gasification and pyrolysis.

The fermentation process can be used to yield hydrogen and organic acids from hexose monomers. Bio-methanol and bio-ethanol can also be converted into H_2.

Biomass gasification produces mostly syngas (CO, H_2) at high temperatures. The main disadvantage of this method is the production of tar and CO_2 released during the decomposition.

Fast pyrolysis follows an endothermic cracking reaction where biomass is converted to bio-oil, char and gas. The bio-oil can be converted to H_2, CO and CO_2 using steam reforming. Yields are considerably lower than with gasification.

Finally, organic waste and waste water can be used as hydrogen resource through a combined anaerobic digestion and catalytic conversion of both CH_4 and NH_3 (obtained from digestate stripping).

16.1.3 Green Ammonia Sources

Due to an ever-increasing need for food and other resources, a large amount of organic waste is produced each year. In 2015, 2 billion tons of waste was generated globally, of which 39% organic waste (Chen et al. 2020). Roughly 780 millions of tons of organic waste is produced annually. Especially sewage sludge and pig, cattle and other types of manure contain high percentages of ammonia, as shown in Tables 16.1 and 16.2.

The occurrence of ammonia in wastes comprises free ammonia (NH_3) and ammonium ions (NH_4^+). Ammonia in waste streams is undesired, but it can be economically extracted. It takes 4.5 mg of oxygen to fully oxidize one mg of ammonia-N. Small concentrations of ammonia can disrupt fauna and flora, leading e.g. to fish kill due to lack of oxygen. In its gaseous form, ammonia is also hazardous for air pollution, potentially forming ammonium aerosols and particulate matter such as

Table 16.1 End and by-products from livestock manure processing in EU (2011)

Type	10^3 t/year
Manure	
Separation solids: Livestock manure solids, typically with a dry matter content of around 25% and rich in phosphorus and nitrogen	4.855
Separation liquids: Liquid fraction, typically with dry matter content around 2% and with a relatively high content of nitrogen and potassium	43,692
Thermal and chemically treated manure: The technologies used in thermal and chemically treated manure are typically standalone technologies	7473
Anaerobic digestion	
Digestate: Product with a lower dry matter and higher mineralized nitrogen content than the undigested raw livestock manure	88.039
Treatment of the solid fraction	
Manure compost: Products with a relatively high content of dry matter, organic matter and nutrients, but with a significant moisture content (i.e. typically >40%, but down to 15%), such as the products of composting	3253
Dried manure and pellets: Low moisture content (<10%) and high concentrations of dry matter, organics and nutrients, such as for thermal drying or pelletizing	967
Ashes and charcoal: High dry matter content but without or with very low organic matter content, such as ashes from combustion or charcoal from pyrolysis. Very low concentration of nitrogen	124
Treatment of the liquid fraction	
Filter water: Water with minimal amount of organic matter, that technically (not legally). Clean enough to be disposed of in the nature, or used as irrigation water;	1732
Manure processing effluents contain some content of organic matter;	6080
Manure concentrates: Minimal amount of organic matter, with a high nitrogen, magnesium or phosphorus fertilizing value;	1154

ammonium nitrate and ammonium sulphate particles that contribute to atmospheric pollution. It can also oxidize in the atmosphere, generating HNO_3, causing acid rain. Ammonia causes a threat not only to the atmosphere or water ecosystems, but also inhibits anaerobic digestion processes by causing a proton imbalance, increasing the energy needed for cellular maintenance, lowering efficiency of enzymatic reactions or changing intercellular pH (Filer et al. 2019; Krakat et al. 2017; Pedizzi et al. 2017). Therefore, removing ammonia from organic waste has 3 main advantages: less water and air pollution, green ammonia production and no inhibitory effect on anaerobic digestion. Palakodeti et al. (2021) reviewed the gas–liquid stripping of ammonia from digestate and found that increasing the temperature, pH, and gas flow rate improved ammonia removal. An increase in temperature leads to a decreased solubility of NH_3 in the liquid phase, an increase in saturation vapor pressure of NH_3 and a shift to the right in the following equation, implying a proposed operation at a high pH:

Table 16.2 N-content and other important concentration data

	DM	N	NH_4-N	P	K	C org	Mg	Ca	Na	Cu	Zn
	%	kg/ton					g/ton				
Separation solids	26.9	7.62	3.68	4.7	3	–	3.31	5.19	0.8	35.5	137.1
Separation liquids	2.86	3.3	3.14	0.46	2.9	63.6	109	37.0	0.67	8.43	24.4
Thermal and chemically treated manure	6.6	2.47	1.18	0.92	2.56	6.62	0.71	–	–	52.3	220
Digestate	5	3.25	2.43	0.73	2.72	42.6	394	2184	954	9	74
Manure compost	46.9	13.9	4.04	6.92	12.6	–	8.97	30.9		58.2	212.5
Dried manure and pellets	87.4	57.8	23.3	22.2	43.3	550	5.54	12.1	7400	183.6	1119
Ashes and charcoal	97.1	19.3	–	38.8	42.8	393	9.75	62.5	13.4	274.3	965
Filter water	0.14	0.2	0.17	0.04	0.24	–	0.16	1.75	0.04	0.01	1.14
Manure processing effluents	3.95	1.51	0.04	1.14	1.94	–	–	–	–	1.51	–
Manure concentrates	6.02	7.25	9.51	0.5	6.52	0	403.7	107	1.57	2.53	12.2

DM dry matter

$$NH_4^+ \leftrightarrow NH_3 + H^+$$

The gas flow rate, does not shift the equation to the right, but rather maintains the driving force for the mass transfer of ammonia.

The ammonia resources, as end and by-products from livestock manure processes and from wastewater treatment sludge are suitable substrates for NH_3 production. Table 16.1 summarize the potential feedstock. Table 16.2 provides composition data.

16.1.4 Literature Data on the Thermal Decomposition of Ammonia

The reversible synthesis and decomposition reaction of ammonia follows the equation:

$$NH_3 \leftrightarrow 0.5\,N_2 + 1.5\,H_2 \qquad \Delta H_0 = 46.2\,\text{kJ/mol}$$

The decomposition reaction is mildly endothermic and needs a catalyst to lower the activation energy. Many possible catalysts can be used for this reaction. The optimal ammonia synthesis catalyst will never be the optimal catalyst for ammonia decomposition, as a result of the very different reaction conditions which result in very different optimal binding energies (Boisen et al. 2005). The rate-determining step depends on whether e.g. Fe, Co and Ni, precious metals (e.g. Ru, Rh, Ir, Pt and Pd), or nonprecious metals (e.g. Ni; Co, Fe). The nitrogen–hydrogen cleavage is considered to be the rate-determining step for precious metals, while N_2 desorption is rate-determining for non-precious metals (Boisen et al. 2005; Jacobsen et al. 2001; Mukherjee et al. 2018). When ranked from the highest to the lowest activity, the following sequence is obtained: Ru > Ni > Rh > Co > Ir > Fe ≫ Pt > Cr > Pd > Cu ≫ Te, Se, Pb (Ganley et al. 2004). The best current catalyst is Ruthenium supported on a carbon carrier. This catalyst will seldom be used because it is very expensive and not readily available (Lendzion-Bielun et al. 2013). Zhang et al. showed promising results with Cobalt-containing carbon nanotubes (CNT) as catalyst for ammonia decomposition, with a nearly 100% conversion obtained at 700 °C. Further research is still required (2007).

Other very good catalysts on a SiO_2 carrier were proposed by Choudhary et al. (2001). Except for a 10% Ni/SiO_2 catalyst with a 70% conversion only, a 65% Ni/SiO_2/Al_2O_3, a 10% Ir/SiO_2 and a 10% Ru/SiO_2 achieved efficiencies between 97 and 98% at temperatures between 650 and 700 °C. As mentioned before, Ru or Ir will not be used because of their high price and limited availability.

Co/Al_2O_3 and Fe/Al_2O_3 also show very good results. A conversion of almost 100% can be obtained, at a lower temperature than needed with CNT's. In the Co/Al_2O_3 catalyst, the active species is Co^0 (Olusola and Sudip 2016). Therefore, a shift from Co^{2+} to Co^0 is desired. Cobalt loadings are normally between 3 and 9 wt% (Bell et al. 2020).

Similar results are found on Fe_3O_4 catalysts supported on γ-Al_2O_3. Cobalt catalysts tend to perform slightly better, but considering the current prices for Cobalt and Iron to be respectively 56,545 USD/ton (Trading economics n.d.) and 162 USD/ton (IndexMundi n.d.), the somewhat better catalytic activity does not outweigh the huge price difference.

Another main concern is choosing an appropriate reactor configuration, as reviewed by Mukherjee et al. (2018). Catalytic membrane reactors show a lot of potential, providing much additional research and improvements are made. A fluidized bed reactor is a good choice when using the Fe_3O_4 supported on γ-Al_2O_3. These Geldart A powder can easily be processed within a fluidized bed reactor.

16.1.5 *Objectives of the Research*

The literature survey reviewed the different H_2 production methods, with special emphases on the catalytic ammonia decomposition. The catalysts were selected. The experimental research will determine hydrogen yields and optimum operating condition. The reaction kinetics will also be determined. Using the experimental results, a scale-up reactor will be designed using Aspen Plus simulation. The layout of the potential plant will be outlined.

16.2 Catalytic Reforming of NH_3

Hydrogen derived from ammonia is extremely pure in comparison with other production ways and no CO and CO_2 in produced.

Large amounts of ammonia are industrially produced around the world and currently the largest part of it is used to produce fertilizers. Many studies however suggest ammonia to be a solution to overcome the current problems regarding H_2 storage and transport, giving ammonia production an even higher demand. Because it is liquid at 25 °C and at a pressure of 8 atm, and has a higher volumetric hydrogen density than liquid hydrogen, ammonia can be easily stored, and thus can be an excellent solution. Therefore, lighter, and smaller transport vessels can be used and liquefaction is less energy intensive. Around 150 Mt of NH_3 is produced per year and transported around the world, using existing marine, road, rail, and pipeline networks. This is an advantage, since this means no new transportation infrastructure needs to be build.

A problem using hydrogen is the low public acceptance regarding safety. But ammonia is already being used by many people and a well-established product. It also has a very recognizable smell, making it easy to detect leakages when the safety systems fail. Ammonia itself can be used as an energy source, but when used for fuel cells, hydrogen is preferred as a reactant.

A scientometric analysis was performed using WoS as Scopus data, imported into VOS viewer software. The importance of the research field is proven by the number of publications, progressively increasing since 2001, including a total of 1047 papers till July 2022. Over the years, the hot research topics have shifted toward, the use of catalysts and nanoparticles. Since 2018, the use of catalyst promoters (bimetallic catalysts, alloy nanoparticles, among others) have been added to the research targets. Ru-, Fe-, Co-, Ce-, Cu-based catalysts are the most used ones, and are well studied since 2015. Latest hot topics are bimetallic catalysts.

16.3 Experimental Investigations

16.3.1 Apparatus and Procedures

For the experimental work, the setup illustrated in Fig. 16.2 was used. Pressurized nitrogen was fed through a flow regulator at 0.3 L/min. The NH_4OH flow was regulated by a positive displacement pump at a flow of 2 mL/min. Nitrogen gas and NH_4OH were fed to an evaporation chamber where NH_3 was released from NH_4OH at 80 °C and pH ~ 9. N_2 and NH_3 were then fed to the catalytic reactor, installed in an isothermal electric furnace (Habertherm). Samples were analyzed at the outlet. Reactor details are given in Fig. 16.3. The catalyst bed is kept in place on both sides by steel fleeces and ceramic fibres, that also acted as a distributor.

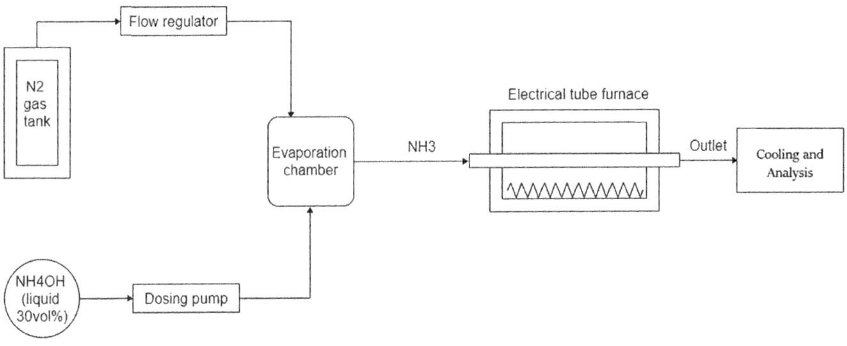

Fig. 16.2 Experimental setup

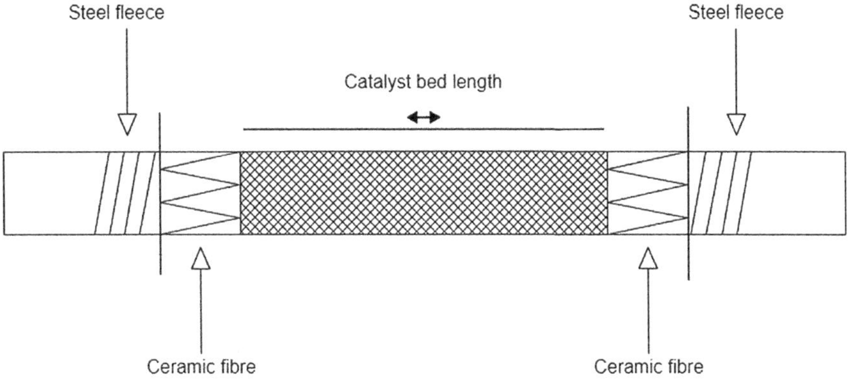

Fig. 16.3 Catalyst reactor in detail

16.3.2 Catalyst Preparation and Properties

The Iron catalyst can be prepared in different ways. In our experiments the wet impregnation method and the dry method were used. A 20 wt% Fe_3O_4 is supported on 80 wt% γ-Al_2O_3.

The wet impregnation method follows Fakeeha et al. (2018). This involves dissolving the Fe-powder in concentrated nitric acid, then bringing it in contact with the γ-Al_2O_3 and stirring at a temperature of 80 °C for 3 h. The catalyst is then dried for 12 h at 120 °C. The catalyst is further calcined in air at 500 °C for 3 h. This catalyst will referred to as Catalyst 1 in the experiments.

The dry method involves adding dispersing ethanol to Fe_3O_4, then adding the alumina and milling it in a bowl for about 10 min. The mixture is dried for 12 h at 120 °C. After the drying, the catalyst is calcined for 3 h at 500 °C. This catalysts is Catalyst 2 in the experiments.

Finally also a Ni/Al_2O_3 catalyst was prepared by the wet impregnation method with a 5 wt% Ni-loading. The same method as for Catalyst 1 was used. NiO was mixed with nitric acid and γ-Al_2O_3. The mixture was stirred at constant rate at a temperature of 80 °C for 3 h. Then the catalyst was dried for 12 h at 120 °C and calcined at 500 °C. This catalyst is Catalyst 3 in the experiments.

After calcination, all 3 catalyst types were re-milled in a Retsch pulverizer.

A BET-scan (Brunauer Emmett Teller) and a particle size measurement were conducted. The BET was determined by Micromeritics Instrument. The particle size and its distribution were measured by Malvern Mastersizer using toluene as the dispersing fluid. Results and comments are summarized in Table 16.3.

In view of the BET values obtained, Catalyst 1 is expected to be more reactive than the other catalysts because of its high BET-value. The Ni-Al_2O_3 catalyst has the lowest BET value (despite the smaller average particle size). Due to this low BET value this catalyst is expected to be the least reactive in the decomposition reaction.

Table 16.3 Catalyst properties

Catalyst	Carrier and method	d_{50} (μm)	BET (m^2/g)	Comments
Fe-Al_2O_3	Wet impregnation	80.60	90.46	The d_{50} corresponds with the d_{50} of the used Al_2O_3 carrier. The active component, Fe, is uniformly distributed in the catalyst pores
Fe-Al_2O_3	Dry milled	86.30	33.45	Milling is inefficient to deposit the catalyst in the pores of the carrier
Ni-Al_2O_3	Wet impregnation	7.49	20.06	Micron-size catalyst and carrier

16.3.3 Experimental Procedure

The following experimental procedure was used. First the temperature of the electric tube furnace was set to a setpoint ranging from 350 to 600 °C. When the setpoint was reached the temperature was kept constant for 10 min and the piston pump was turned on. The pump was set to pump 2 mL/min of NH_4OH into the evaporation chamber. During all the experiments carrier N_2-gas was led through the evaporation chamber at a rate of 0.3 L/min. The reaction gases were released and a sample was continuously analyzed by GC–MS (gas chromatography-mass spectrometry). After repeat analysis, a new temperature could be set and the process was repeated.

16.3.4 Hydrogen Yield for Different Operating Conditions (T, Type of Catalyst)

The results are presented for each catalyst at different temperatures. Figure 16.4 presents the results as the NH_3 outlet concentration (in mol/min) plotted against temperature (in K). The initial concentration of ammonia (Co) is indicated.

16.3.5 Kinetics of the NH_3 Decomposition Reaction

Kinetic Equations. For reactants to be transformed into products, chemical bonds must be rearranged mostly due to rupture of some of the bonds, which occurs when molecules possess an energy level equal to or in excess of the activation energy (E_a). Only then can products be formed, as illustrated in Fig. 16.5 for endothermic reactions.

Heat is often the source of the activation energy in endothermic reactions. The temperature of a system affects the average kinetic energy of all the atoms and

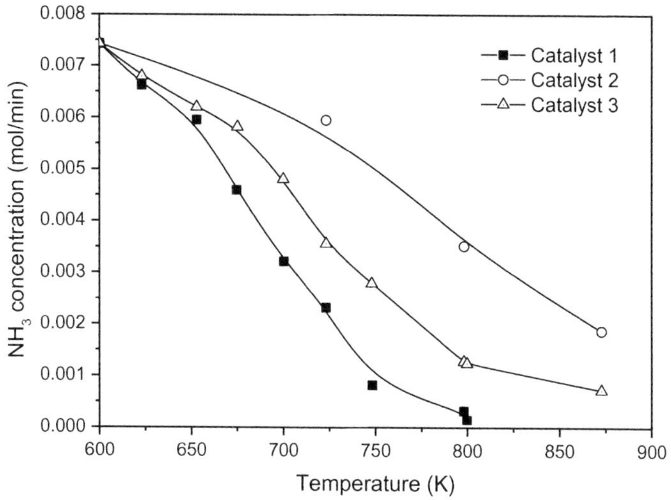

Fig. 16.4 NH_3 concentration versus temperature (Co ~ 7.4 mmol/min)

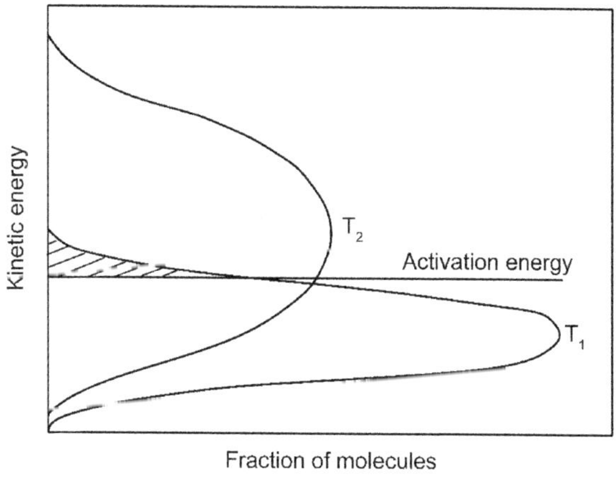

Fig. 16.5 Molecular kinetic energy distribution at 2 different temperatures ($T_2 > T_1$)

molecules present in the system, with higher temperatures increasing the average kinetic energy and broadening the distribution of energies, as illustrated in Fig. 16.5. The concentration of suitably activated reactant molecules will therefore increase with temperature and the energy level of an increasing fraction of the molecules will exceed the required reaction activation energy.

The Arrhenius equation expresses how the reaction rate and the temperature are related:

$$k_r = A\exp\left(-\frac{E_a}{RT}\right) \tag{16.1}$$

The exponential nature of the relationship explains why s mall changes of E_a and of T will considerably affect the reaction rate. The Arrhenius pre-exponential factor, A, is a probabilistic factor to accommodate supplementary requirements that affect the reaction rate such as specific catalyst properties (e.g. type of catalyst, wash coat crystal structure, number of active sites). Variations of A are also the result of the catalyst deactivation, such as if catalyst coking occurs.

The exponential nature of the relationship explains why s mall changes of E_a and of T will considerably affect the reaction rate. The Arrhenius pre-exponential factor, A, is a probabilistic factor to accommodate supplementary requirements that affect the reaction rate such as specific catalyst properties (e.g. type of catalyst, wash coat crystal structure, number of active sites). Variations of A are also the result of the catalyst deactivation, such as if catalyst coking occurs.

Equation (16.2) demonstrate that the apparent reaction rate constant k_0 combines the intrinsic reaction rate constant, k_r and the mass transfer coefficient k_m. Smolders and Baeyens (2004) demonstrated that k_0 depends on the operating temperature according to:

$$k_0 = \frac{Ak_m \exp(-\frac{E_a}{RT})}{k_m + \exp(-\frac{E_a}{RT})} \tag{16.2}$$

with k_m only slightly dependent of temperature. This is illustrated in Fig. 16.6.

At low temperatures: $k_m \gg A\exp(-\frac{E_a}{RT})$ or $k_0 = A\exp\left(-\frac{E_a}{RT}\right)$, meaning that the external mass transfer is faster than the chemical reaction, and the overall reaction kinetics is controlled by the chemical regime. A linear plot of (ln k_0) versus (1/T), as illustrated in Fig. 16.6, allows to determine the Arrhenius parameters *Ea* and A.

At high temperatures: $k_m \ll A\exp(-\frac{E_a}{RT})$, thus low 1/T, the overall reaction kinetics proceed in the mass transfer (diffusion) controlled regime. Since the mass transfer coefficient k_m is a minor function of temperature within the operating high temperature range, the slope of ln k_0 is less pronounced as shown in Fig. 16.6.

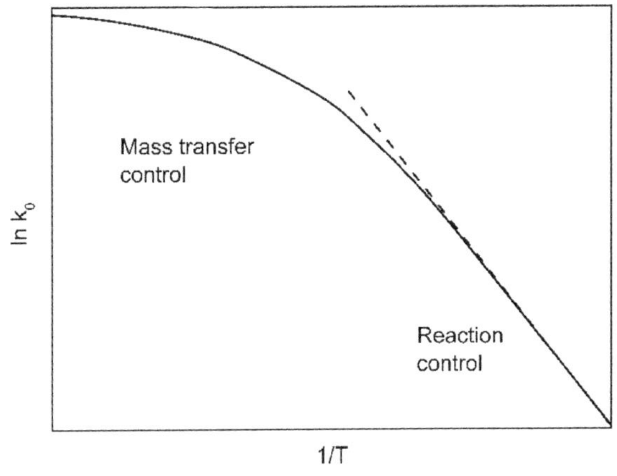

Fig. 16.6 Arrhenius plot in the chemical (higher 1/T) and in the diffusion-controlled (lower 1/T) regions

Expressing the Rate Constant k_r to Include the Catalyst Geometry. Although the kinetics is mostly defined by the reaction under security, the specific characteristics of the catalyst itself, such as weight, specific surface area, fractional coverage, and wash coat thickness are important. More general kinetic expressions will include the catalyst properties, commonly performed by introducing parameters such as the area velocity (AV in $m^3/(m^2\ h)$), the specific catalyst surface area (AP in m^2/m^3), and the space velocity (SV in h^{-1}).

Combining these catalyst characteristics with the reaction rate constant introduces a catalyst independent kinetic constant, K.

$$K = \frac{3600 \cdot k_r}{AP} = \frac{3600 \cdot k_r \cdot AV}{AP}(\mathrm{m}^3/(\mathrm{m}^2\mathrm{h})) \tag{16.3}$$

These parameters are determined, either by measured BET-value (m^2/g), by the weight and volume of the catalyst present (g), by the fractional coverage of the carrier, f_V, and the thickness of the wash coat (m).

Kinetic Results. Since the initial ammonia concentration and the output concentration of ammonia are available, the conversion can be determined, by subtracting the output from the initial concentration, and divided by the initial concentration. The resulting curves are illustrated in Fig. 16.7.

Then the residence time needs to be calculated. This is dependent on the volume of the catalyst as well as the flow rate and temperature. First the residence time at 20 °C for each catalyst is calculated using Eq. (16.3), then the residence times are converted to the specific temperature using a correction factor, A.

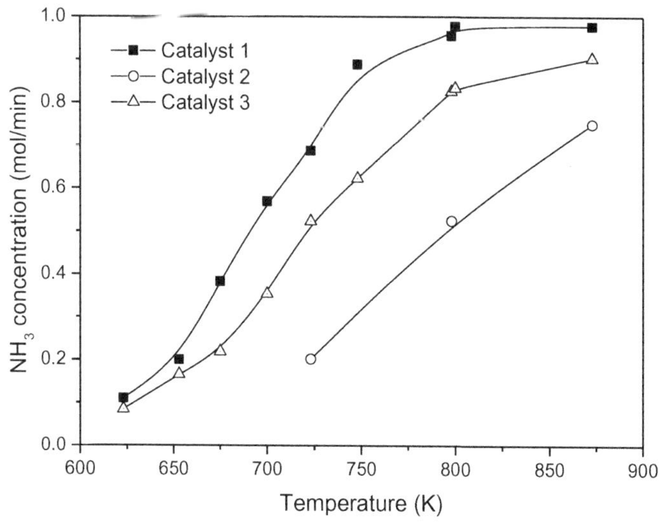

Fig. 16.7 NH_3 conversion versus temperature

$$t = \frac{V(m^3)}{F\left(\frac{m^3}{s}\right)} \tag{16.4}$$

With $A = \frac{273+20}{273+T}(T : temperature\, in, \,°C)$.

Since the residence times (t) and corresponding conversions (η) are known, the reaction rate constant (k) can be derived, using Eq. (16.5).

$$\ln(1 - \eta) = -k * t \tag{16.5}$$

When ln(k) is plotted versus 1/T a straight line, indicating a linear relationship, should be required. The plot is displayed in Fig. 16.8. This linearity indicates that the reaction mechanism is purely reaction limited within the experimental T-range. E_a and A can be found based on Arrhenius Eq. (16.1).

The resulting activation energy and maximum reaction speed are presented in Table 16.4.

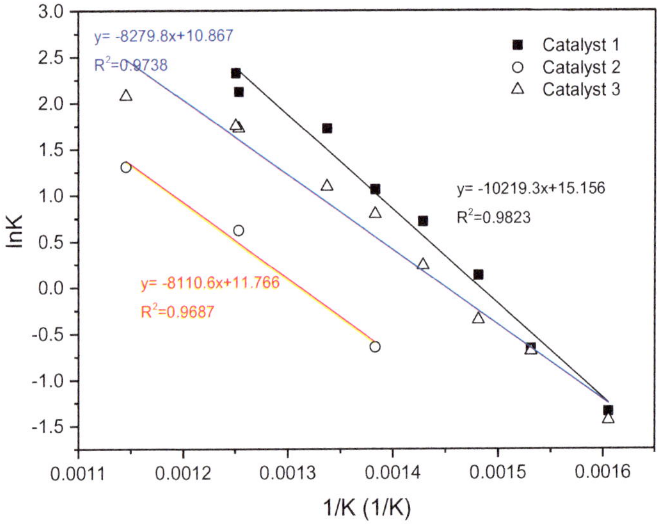

Fig. 16.8 Arrhenius

Table 16.4 Activation energy and maximum reaction speed per catalyst type

Catalyst type	Slope	Ea	ln(A)	A
Cat 1	−10,219	84,960.77	15.16	3,820,913
Cat 2	−8279	68,831.61	10.87	52,417.72
Cat 3	−8110.6	67,431.53	11.77	128,797.9

Additionally, this data can be used to calculate the conversions at different residence times and temperatures. In Fig. 16.9 the conversion is plotted versus the residence time at 3 different temperatures: 723, 798 and 873 K. In Fig. 16.10 the conversion is plotted versus temperature at 3 different residence times: 0.2, 0.3 and 0.4 s.

From these plots can be deduced that Catalyst 1 is by far the best suited catalyst for this reaction, followed by Catalyst 3. Catalyst 2 performs remarkably worse then the other 2. This indicates that the dry milling manufacturing is less appropriate. Catalyst 1 and 3 are very suitable for the ammonia decomposition reaction and can be used in upscaling processes.

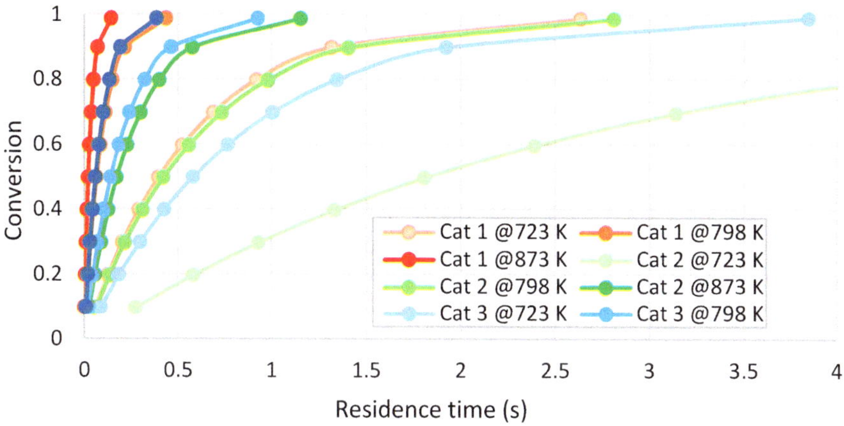

Fig. 16.9 Conversion versus residence time at different temperatures

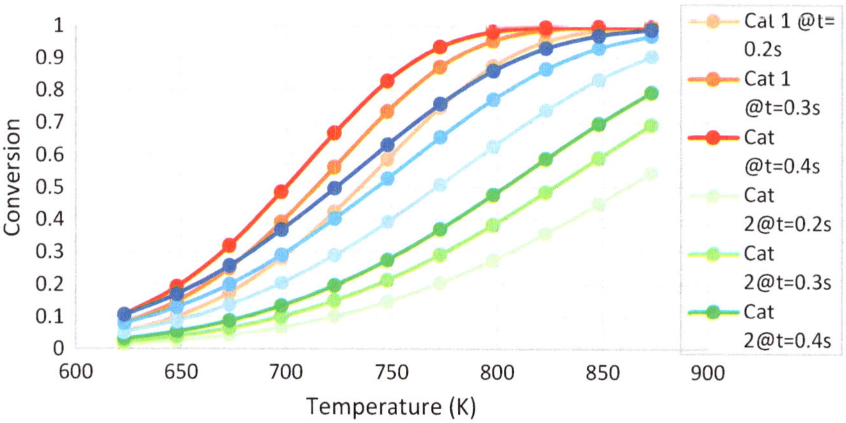

Fig. 16.10 Conversion versus temperature at different residence times

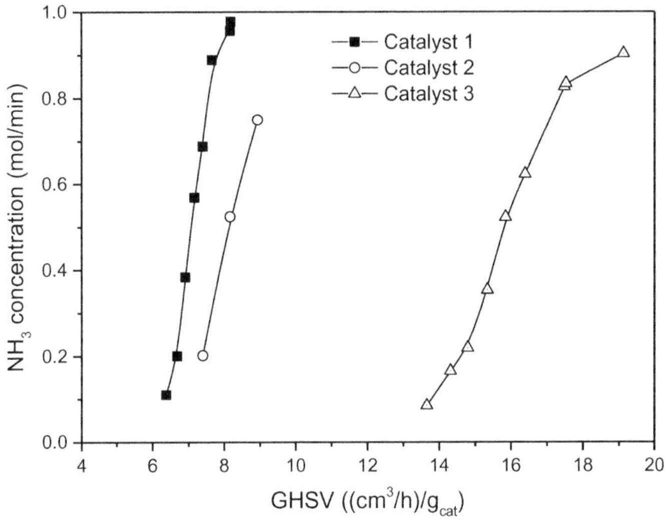

Fig. 16.11 Conversion versus GHSV

Additionally, a last parameter, the GHSV (gas hourly space velocity in $(cm^3/h)/g_{cat}$)) can be calculated using Formula 16.6.

$$GHSV = \frac{F\left(\frac{cm^3}{\min}\right)}{m_{cat}} * 60 \tag{16.6}$$

In Fig. 16.11, the conversion is plotted versus the GHSV. The GHSV is much higher for catalyst 3 because the density of the Ni-catalyst is much lower than the density of the Fe-catalyst, leading to a higher GHSV.

16.4 Recommendations Toward Scale-Up

16.4.1 Design of a Pilot-Scale Reactor

From the experimental results, the Fe/Al_2O_3 catalyst was selected. The NH_3 conversion is close to 100% at 500 °C and the kinetics are fast, with a rate constant of about 7.5 s^{-1}. It is expected that about 5000 ton of NH_3 will be treated per year. The capacity of the design was hence fixed at 600 kg/hr, leading 106.5 kg/hr of hydrogen. The plant will not operate on a 24/24, 7/7 basis. One month downtime is expected. The annual production of hydrogen could therefore be estimated at about 844 ton/year. A fluidized bed is selected as reactor for its high conversion efficiency and isothermal operation. Aspen V.12 software was used for the design.

16.4.2 Design Procedure

The catalyst is characterized as group A powder, and hence suitable for turbulent fluidization (Deng et al. 2021). The fluidization velocity is chosen at 0.3 m/s. A freeboard (expanded bed section) will limit the carry-over of the catalyst. A freeboard cyclone with dip-leg is also installed.

The Aspen flowsheet of the process is illustrated in Fig. 16.12. Ammonia from digestate will be fed to a compressor (COMP) to feed the reactor at 1.5 bar, thus accounting for the pressure drop over the fluidized bed, filter, separator, and other equipment. The fluidized bed reactor (FLUIDBED) will operate at 500 °C. The catalyst bed has a bed mass of 4370 kg. The diameter widens from 1 m at the bottom to 3 m at the top and has a height of 5 m in total. The gas stream (N_2 and H_2) (GASOUT) will leave the reactor and go through a high temperature filter (HTFIL TER). Here the remaining particles (SOLIDOUT) will be removed and send to the mixer (MIX) before being reinjected into the fluidized bed reactor. The gas stream leaving the filter (GAS) will be enriched in H_2 by a membrane (SEPARAT). The N2OUT stream contains mostly N_2, with traces of NH_3. The exhaust steam can be released in the environment. The internal cyclone will recirculate catalyst to the mixer (CATRECYC).

The NH_3 distributor is of a perforated plate type with 6000 orifices a pressure drop of 0.045 bar. The high-temperature filter has a total filtering area of 50 m^2 and a maximum pressure drop of 0.05 bar. For candles of 0.05 m internal diameter and 3 m length, a minimum of 106 candles is required. N_2 gas is used for pulse jet cleaning, at a jet velocity of 0.15 m/s to maximally remove the cake layer (Li et al. 2021b).

The simulation results provided information about the geometry of the used equipment, energy requirements, and prevailing process conditions. Paired with this yield an isentropic compressor of 14.5 kW will be required. The reactor will have a much higher energy consumption with an endothermic heat duty of 712 kW. Heat recovery on the exhaust gas will be applied.

16.4.3 Design Results

The Aspen Plus input and output data are summarized in the Table 16.5.

16.4.4 Important Considerations

Storage of Hydrogen. A production rate of 106.5 kg/hr of hydrogen leads to volume flow of 3000 m^3/hr at 500 °C. Since the heat of the produced H_2 and N_2 are used to preheat the ammonia and N_2, the volume of can be reduced from 3000 m^3/hr to 1280 m^3/hr at 25 °C.

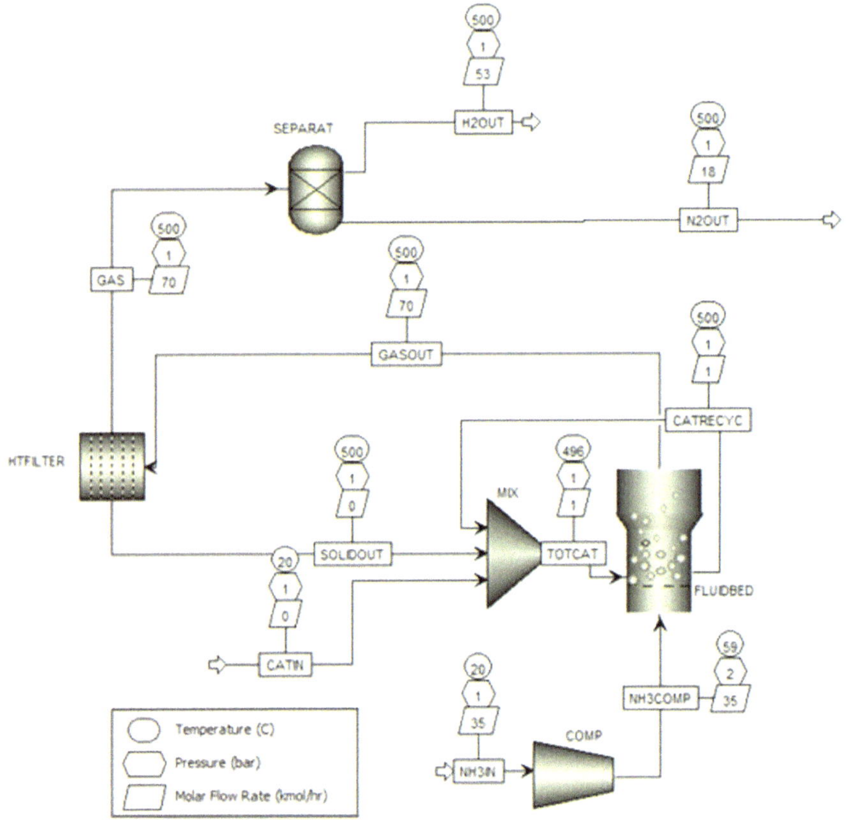

Fig. 16.12 ASPEN flowsheet

To limit the volume of the storage tanks, and unless a continuous use of the H2 can be realized (as feed with biogas to the CHP generators, or as feed gas into the anaerobic digestion to increase the CH4 yield) storage and/or compression of H2 is proposed. This will add to the costs (investment and operation) but is deemed necessary. Several methods were assessed by Andersson and Gronkuist (2019). Absorption of H_2 into solid materials and metal hydrides seem the most feasible materials at the moment for this type of storage. If the time of storage of H2 remains low, the pressurized storage is the best choice from an economical point of view. It is not clear yet if the solid storage methods will be economically more feasible.

A buffer storage period of 4 h is selected. From the production rates, the required storage capacity can then be derived. The discharge temperature of the compressor is assumed to have the same temperature as its inlet. Since it is a three-stage compressor with intercooling, this is not an unrealistic assumption. At 50 bar, storage volume of 240 m^3 will still be required.

Table 16.5 Aspen input and output data

Compressor data			
Discharge pressure (bar)	1.5	Power (kW)	13.5
Inlet temperature (°C)	20	Efficiency (%)	0.72
Type	Isentropic compressor	Outlet temperature (°C)	59
Fluidized bed reactor input data		*Output data*	
Bed weight (kg)	4370	Height of bottom zone (m)	1.33
Operation temperature (°C)	500	Height of freeboard (m)	3.66
Height (m)	5	Overall pressure drop (bar)	0.33
Bed diameter (m)	1	Heat duty (kW)	711.8
Freeboard diameter (m)	3		
Type of distributor	Perforated plate		
Distributor pressure drop (bar)	0.045		

Safety Considerations. Combined with oxygen and other elements, hydrogen has a wide range of explosion limits in air and is very flammable. It is moreover very prone to leakages (30–180% larger than methane and four times larger than air) due to its small molar volume. Hydrogen has however no toxic properties. Specific safety measures must be accounted for, although the fire damage caused by hydrogen flames is less severe than the same amount of hydrocarbons on fire would cause. In addition the burning time is a factor 10 lower.

16.5 Conclusions

The production of H_2 from "green" or industrial NH_3 was studied in the present research. The mildly endothermic decomposition of NH_3 into H_2 and N_2 is achieved with a 100% yield at 500 °C using a Fe-γ Al_2O_3 catalyst. The alternative and more expensive Ni-γ Al_2O_3 catalyst scores significantly lower. Experimental results were converted into reaction kinetics. The reaction is very fast, with a rate constant of 7.5/ s at 500 °C. Full conversion is obtained within 0.4 s. The experimental data were further used in the Aspen-based design of a ~850 T/year H_2 production facility. Very pure (>95%) H_2 can be produced and can be stored by several techniques. It is proven that the H_2 production from NH_3 has a high potential, and merits pilot-scale consideration.

References

Andersson J, Grönkvist S (2019) Large-scale storage of hydrogen. Int J Hydrog Energy 44:11901–11919

Bell TE, Ménard H, González Carballo J-M, Tooze R, Torrente-Murciano L (2020) Hydrogen production from ammonia decomposition using Co/γ-Al_2O_3 catalysts—insights into the effect of synthetic method. Int J Hydrog Energy 45:27210–27220

Boisen A, Dahl S, Norskov J, Christensen C (2005) Why the optimal ammonia synthesis catalyst is not the optimal ammonia decomposition catalyst. J Catal 230:309–312

Chen DM-C, Bodirsky BL, Krueger T, Mishra A, Popp A (2020) The world's growing municipal solid waste: trends and impacts. Environ Res Lett 15:074021

Choudhary TV, Sivadinarayana C, Goodman DW (2001) Catalytic ammonia decomposition: COx-free hydrogen production for fuel cell applications. Catal Lett 72:197–201

Deng Y, Sabatier F, Dewil R, Flamant G, Le Gal A, Gueguen R, Baeyens J, Li S, Ansart R (2021) Dense upflow fluidized bed (DUFB) solar receivers of high aspect ratio: different fluidization modes through inserting bubble rupture promoters. Chem Eng J 418:129376

Deng Y, Dewil R, Appels L, Van Tulden F, Li S, Yang M, Baeyens J (2022) Hydrogen-enriched natural gas in a decarbonization perspective. Fuel 318:123680

Deng Y, Liu J, Li S, Dewil R, Zhang H, Baeyens J, Mikulčić H (2022) The steam-assisted calcination of limestone and dolomite for energy savings and to foster solar calcination processes. J Clean Prod 132640

Filer J, Ding HH, Chang S (2019) Biochemical methane potential (BMP) assay method for anaerobic digestion research. Water 11:921

Foged HL, Flotats X, Bonmatí A, Palatsi J (2011) End and by-products from livestock manure processing—general types. Chem Compos Fertil Qual Feasibility Market (2011)

Fakeeha AH, Ibrahim AA, Khan WU, Seshan K, Al Otaibi RL, Al-Fatesh AS (2018) Hydrogen production via catalytic methane decomposition over alumina supported iron catalyst. Arab J Chem 11:405–414

Ganley JC, Thomas FS, Seebauer EG, Masel RI (2004) A priori catalytic activity correlations: the difficult case of hydrogen production from ammonia. Catal Lett 96:117–122

IEA (2019) The future of hydrogen

Jacobsen CJH, Dahl S, Clausen BS, Bahn S, Logadottir A, Nørskov JK (2001) Catalyst design by interpolation in the periodic table: bimetallic ammonia synthesis catalysts. J Am Chem Soc 123:8404–8405

Krakat N, Demirel B, Anjum R, Dietz D (2017) Methods of ammonia removal in anaerobic digestion: a review. Water Sci Technol 76:1925–1938

Lendzion-Bielun Z, Narkiewicz U, Arabczyk W (2013) Cobalt-based catalysts for ammonia decomposition. Materials (Basel) 6:2400–2409

Li S, Zhang H, Nie J, Dewil R, Baeyens J, Deng Y (2021a) The direct reduction of iron ore with hydrogen. Sustainability 13:8866

Li S, Baeyens J, Dewil R, Appels L, Zhang H, Deng Y (2021b) Advances in rigid porous high temperature filters. Renew Sustain Energy Rev 139:110713

Mukherjee S, Devaguptapu SV, Sviripa A, Lund CRF, Wu G (2018) Low-temperature ammonia decomposition catalysts for hydrogen generation. Appl Catal B Environ 226:162–181

Olusola OJ, Sudip M (2016) Temperature programme reduction (TPR) studies of cobalt phases in -alumina supported cobalt catalysts. J Pet Technol Altern Fuels 7:1–12

Palakodeti A, Azman S, Rossi B, Dewil R, Appels L (2021) A critical review of ammonia recovery from anaerobic digestate of organic wastes via stripping. Renew Sustain Energy Rev 143:110903
Pekic S (2021) Facts and factors: green hydrogen market to grow to $1.423m by 2026
Pedizzi C, Lema JM, Carballa M (2017) Enhancing thermophilic co-digestion of nitrogen-rich substrates by air side-stream stripping. Bioresour Technol 241:397–405
Smolders K, Baeyens J (2004) Thermal degradation of PMMA in fluidised beds. Waste Manag 24:849–857
Zhang J, Comotti M, Schüth F, Schlögl R, Su DS (2007) Commercial Fe- or Co-containing carbon nanotubes as catalysts for NH 3 decomposition. Chem Commun 1916–1918

Chapter 17
Concentrated Particle-Driven Solar Power Receiver: Experimental and Simulation Hydrodynamic and Thermal Characteristics

Yimin Deng, Andrés Reyes Urrutia, Maarten Vanierschot, Jan Baeyens, and Germán Mazza

Abstract Concentrated solar power plants are increasingly developed, and particle-driven concepts offer a high potential in the receiver and heat storage. Both the high temperature effects in solar receivers, and the use of Computational Fluid Dynamics (CFD) have been largely ignored in trying to elucidate their hydrodynamic and thermal behavior. The present paper assesses solid and gas flow modes, at different operating conditions. Wall-slugs are always present in the tube, especially in its hot section. Solids back mixing between the hot and cold zones is important and alters the temperature profile along the bed height. Axial solids and gas velocity profiles demonstrate the existence of three flow modes. In the center or core of the column, occupying 50–60% of its diameter, the bubbles induce an upward movement of both phases. In most of the annulus area outside the core zone, both phases flow downwards in a bubble-lean zone. In a very thin zone adjacent to the wall, the mean particle and gas velocities are positive (upward flow) due to the occurrence of wall-slugs.

Y. Deng
Department of Chemical Engineering, Environmental and Process Technology Lab., KU Leuven, 2860 Heverlee, Sint-Katelijne-Waver, Belgium

A. R. Urrutia · G. Mazza (✉)
PROBIEN - Instituto de Investigación y Desarrollo en Ingeniería de Procesos, Biotecnología y Energías Alternativas (Universidad Nacional del Comahue - CONICET), z1400, CP8300 Neuquén, Argentina
e-mail: german.mazza@probien.gob.ar

M. Vanierschot
Department of Mechanical Engineering, KU Leuven, Group T Leuven Campus, Celestijnenlaan 300, 3001 Heverlee, Belgium

J. Baeyens (✉)
Beijing Advanced Innovation Center for Soft Matter Science and Engineering, Beijing University of Chemical Technology, Beijing 100029, China
e-mail: baeyens.j@buct.edu.cn

J. Baeyens et al. (eds.), *Proceedings of 2022 4th International Conference on Environment Sciences and Renewable Energy*, Environmental Science and Engineering, https://doi.org/10.1007/978-981-19-9440-1_17

Keywords Concentrated solar power receiver · Particle suspension · CFD modeling · Geldart A particles · Small diameter fluidized bed · Wall-slugging

17.1 Introduction

The use of small I.D. beds is normally limited to laboratory research, although it also represents the individual bed cells between vertical heat exchanger tubes in a catalytic reactor (Deng et al. 2021; Zhang et al. 2021). Novel developments of particle-in-tube solar heat receivers however use such small I.D. beds (Zhang et al. 2016). This novel concept is referred to as particle-in-tube or Upflow Bubbling Fluidized Bed (UBFB) system, and used in the on-sun testing of a single tube (36 mm I.D., 2.6 m long but with only a 0.5 m long section subjected to the concentrated solar beam) and of a 150 kW pilot module consisting of 16 parallel tubes of 29.7 mm I.D., each 1 m long. Geldart A SiC particles were fluidized at superficial gas velocities of ~0.03 to 0.25 m s^{-1} (Zhang et al. 2017).

17.1.1 A-Type Powders and Their Fluidization Behavior in Small I.D. Fluidized Beds

Previous investigations studied the fluidization of Geldart A and B particles, at a superficial velocity exceeding the onset of slugging. They report the formation of axi-symmetric slugs for any column diameter (Baeyens and Geldart 1974; Kong et al. 2017). In Geldart A particles, bubbles occur at gas velocities in excess of U_{mb}, the minimum bubbling velocity (Kong et al. 2017). A further increase in gas velocity will lead to a significant bed expansion and a gross circulation of the powders occurs as soon as bubbles are present, producing considerable particle mixing. Bubbles in group A powders split and recoalesce frequently, resulting in a maximum stable bubble size if the diameter of the bed is big enough to avoid slugging (Kong et al. 2017). At a sufficiently high superficial gas velocity, and in tall beds of small I.D., bubbles will transform into axi-symmetric or wall-slugs (Baeyens and Geldart 1974; Kong et al. 2017). Slugging minimizes the gas of by-passing solids, and significantly hampers the solids mixing and wall-to-bed heat transfer potential. The prediction of the height of transition from freely bubbling to slugging conditions is hence of paramount importance, and currently only known from experimental investigations where the onset of slugging is shown to mostly depend upon the tube diameter and the excess velocity. Kong et al. (2017) observed that for moderate to high excess gas velocities, the fluidization along tall beds may be divided into three different regions: a bottom zone, close to the distributor, with incipient bubbling; a narrow zone of bubble growth and coalescence; and finally, a slugging region that prevailed through/along the tube height until slugs erupt at the bed surface. Axi-symmetric slugging was found to occur from a bed height exceeding 1.5 m in a 4 m tall fluidized bed of 0.05 m I.D. (Kong et al. 2017).

The reduced heat transfer was previously studied by Reyes Urrutia et al. (Kong et al. 2017). Their simulations also demonstrated distinct particle circulation patterns along the height of the fluidized bed column, with an upward core and a downward annulus region. Wall-slug formation along the tube was also predicted by the simulations. Slugs significantly reduce both the bubble-induced particle mixing and the associated heat transfer. It is hence important to investigate the bubbling/slugging effects in a geometry similar to the currently developed receiver concept.

17.1.2 Objectives

Apart from the recent study by Kong et al. (2017) and the initial study by Baeyens and Geldart (1974), the assessment of the behavior of A-type powders in tall tubes of small I.D. has not been studied extensively. These studies are moreover limited to ambient temperature, thus ignoring operations at higher temperatures such us experienced in solar receivers or catalytic reactors with embedded vertical heat exchangers. These systems were only scarcely been evaluated by CFD modeling.

The present paper attempts to provide answers to the different prevailing issues (i) by considering the upflow behavior of Geldart A particles in a tall but small I.D. tube; (ii) by determining the prevailing hydrodynamics, with different solid and gas flow regimes occurring for different operating conditions of gas and solid flow rate, and this at ambient or high temperature; (iii) by introducing the most appropriate predictions of powder-gas properties for fluidization of Geldart A particles, (iv) by applying a commercial simulation in ANSYS-Fluent 17 using the theory of granular flow, and using a structured mesh, refined in the region near the tube-wall; and finally (v) by validating the approach through comparing simulation and experimental results. The research involved a stepwise approach, as illustrated in Fig. 17.1.

17.2 Modeling and Simulation

Simulations have been performed with the ANSYS-Fluent code (Engineering Simulation and Scientific Software—ESSS—Academic Research) version 17.

17.2.1 Physical and Geometrical Parameters

The geometry adopted for the simulations corresponded to the experimental setup, and used a tube 0.036 m I.D. and 2.6 m long, with a heated zone of 0.5 m long. The Geldart A particles used were silicon carbide of average particle size 64 μm,

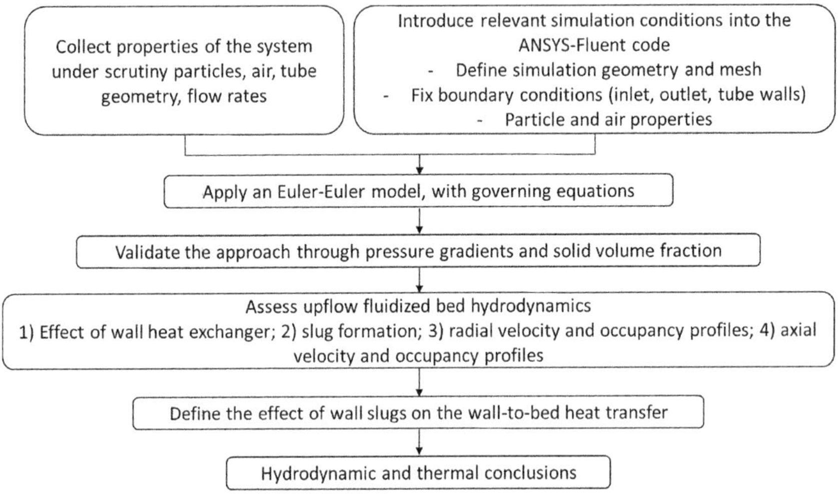

Fig. 17.1 Logic diagram of the objectives and of the layout of the paper

sphericity 0.77 and absolute particle density 3210 kg/m^3 (Kong et al. 2019). Thermal properties (heat capacity and thermal conductivity) of SiC particles are expressed as polynomial functions of the temperature, using the NIST database. Particle emissivity was obtained for relevant operation temperatures and only slightly decreased from 0.86 (400 K) to 0.85 (1200 K). Absorptivity data were taken from Kuipers et al. (1992). U_{mf}, U_{mb}, ε_{mf} and ε_{mb} were experimentally determined as 5 and 6.6 mm/s, and 0.57 and 0.59 respectively. The tube was insulated from a height of 0 to 0.8 m, and from 1.3 to 2.6 m. Only the surface between 0.8 and 1.3 m was subjected to solar irradiation. The velocity at height 0 m is fixed. The pressure at the top is atmospheric.

A cold simulation (without heating) was carried out as reference to enable a further comparison with hot experiments and show the effects of temperature on particle flow and heat transfer. In this case, it was not necessary to solve the energy balance.

The ambient temperature simulation (Run 1, without considering wall heat transfer) was performed with 87.8 kg/h SiC mass flow rate. The hot simulations (under solar irradiance) were carried out at solid mass flow rates of 87.8 kg/h (Run 2) and 42.4 kg/h (Run 3). The applied superficial air velocities were 0.039 m/s (for Run 1 and Run 2) and 0.025 m/s (for Run 3). Table 17.1 summarizes the essential operating parameters and boundary conditions of the three runs. Air expansion at higher temperatures was accounted for by means of ANSYS-Fluent 17 routines. Experiments were previously described in detail (Zhang et al. 2016, 2017).

Table 17.1 Operating parameters, boundary conditions used in CFD simulations and experimental values of $T_{p,i}$, $T_{p,o}$ and ε_p

	Run 1	Run 2	Run 3
$\dot{m}_{SiC}$ (kgh^{-1})	87.80	87.80	42.35
$T_{p,cold}$ (K)	–	317.28	300.53
$T_{p,i}$ (K)	–	389.62	455.78
$T_{p,o}$ (K)	–	495.83	548.83
$T_{w,z}^{int,i}$ (K)	–	504.83	556.17
$T_{w,z}^{int,m}$ (K)	–	560.62	604.41
$T_{w,z}^{int,o}$ (K)	–	561.08	606.62
ε_p	0.31	0.31	0.31
u_p $[m\,s^{-1}]$	0.024	0.024	0.012
u_g $[m\,s^{-1}]$	0.039	0.039	0.025
$T_{p,cold} = T_{g,cold}$ \; [K]	298.15	317.28	300.53
Interval	–	$0.8m < z \le 1.05m$ $1.05m < z < 1.3m$	$0.8m < z \le 1.05m$ $1.05m < z < 1.3m$
$T_{w,z}^{int}$[K]	–	$401.8 + 196.92z$ $616.97 - 7.96z$	$326.3 + 223.16z$ $558.69 + 1.84z$

17.2.2 Simulation Mesh

The discretization of the domain was carried out using a structured mesh, refined in the region near the wall of the tube. It consists of a total of 920,400 hexahedral elements (Fig. 17.2). During the grid validation procedure, it was verified that the grid was fine enough to provide grid-independent results. To this end, three different variables were used as convergence criteria: pressure drop, solid temperature values at the inlet of the heating zone and also at the outlet of the heat exchange zone.

17.2.3 Boundary and Initial Conditions

Base of the Tube, Inlet Conditions. Energy: Air and SiC particles enter at an average temperature, as experimentally determined, $T_{p,cold}$.

Momentum: Solid enters the tube, with upward linear velocity and volume fraction ε_p, corresponding to the average values determined from the solid flow rate and the pressure drop of each experimental run. For the gas phase, the interstitial velocity is $u_g = u_p + u_{mf}$. The velocity direction of both phases is supposed to be perpendicular to the entry plane. Operating conductions for each stage are shown in Table 17.1.

Tube Walls. Energy: For simulations with heat exchange in the 0.5 m high region of the tube exposed to concentrated solar energy, a temperature profile was set along

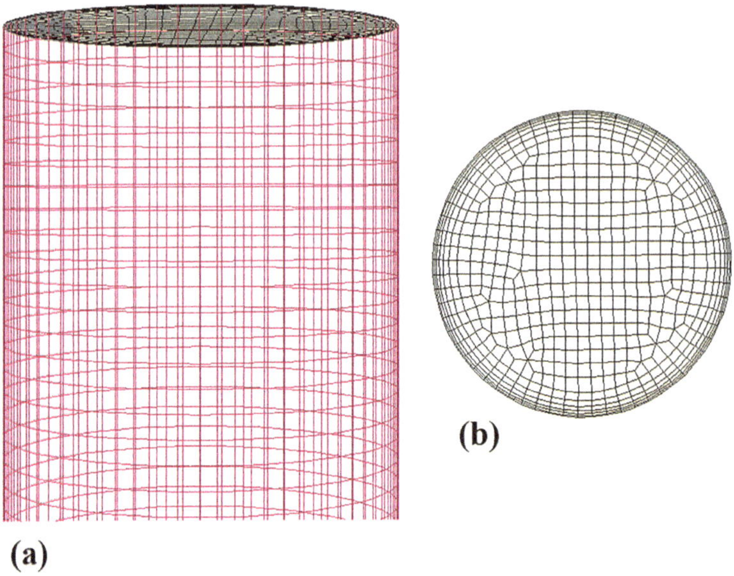

Fig. 17.2 Structured mesh in the tube used in simulations. **a** Side view and **b** cross-sectional view

the z-coordinate, defined from the experimentally determined average wall temperature. The internal wall temperature as a function of height (z) was programmed in C++ language and incorporated into the simulation algorithm, according to the functions "Interval" and $T_{w,z}^{int}$[K] of Table 17.1. The walls of the regions before and after the heated zone were considered adiabatic.

Momentum: the no-slip condition along the wall was applied to both phases throughout the simulations.

Top of the Tube, Outlet Condition. Since the gas was exhausted to the atmosphere, the gauge pressure was set at zero. Only gas flow perpendicular to the surface of the exit plane was allowed.

Initial Conditions. The simulations started with a fixed bed of SiC particles, of 0.2 m height, and a voidage $\varepsilon_g = 0.4$. For simulations with heat exchange, the initial system temperature is that of the emulsion at the bottom ($T_{p,cold}$).

17.2.4 Numerical Modeling of Multiphase Flow

The chosen multiphase model is the Euler-Euler model, widely used in the modeling of fluidized systems (Kuipers et al. 1992; Taghipour et al. 2005; Armstrong et al. 2010; Gidaspow et al. 1992). The model-related equations are summarized in Appendix, where Table 17.4, summarizes the equations for both air and solid phases. As already mentioned, particle velocity variations were generated from collisions and quantified

Table 17.2 Main numerical parameter applied in simulations

Particle diameter	64 μm
Restitution coefficient	0.9
Time step	0.001 s
Time discretization	2nd order implicit
Iterations per step	40
Drag model	Syamlal and O'Brien (1987)
Maximum error for convergence	0.0001

by ANSYS-Fluent 17 using the kinetic theory of granular flow (KTGF). In order to solve the momentum exchange between the phases, the Syamlal and O'Brien drag model (1987) was used. In Table 17.5, the coefficients C1 and C2, characteristic of the drag model, were determined from the values of velocity and porosity in the condition of minimum fluidization of the solid–gas pair being studied.

For the resolution of the convective heat exchange between phases, the Nusselt number was calculated from the correlation of Gunn (1978), applicable in a wide range of porosities and Reynolds numbers ($\leq 10^5$).

The continuous intrinsic nature of the Euler-Euler model generates an overestimation of the energy transferred between the wall and the dense particle emulsion (Gidaspow et al. 1992). For the correct resolution of the wall-emulsion heat transfer, it was necessary to incorporate a model of effective conductivity for both phases. Table 17.5 also shows the Zehnder and Schlünder model, previously demonstrated by Kang et al. (2019) as fitting experimental data with good accuracy.

All simulations were carried out in transient state, adopting a time step of 0.001 s with 40 iterations per step, which ensured the adequate convergence during the simulations. The pressure-based resolution method was selected. A second order scheme was used for the spatial discretization of the convective terms in the conservation equations. Table 17.2 summarized the values of main numerical parameters and conditions adopted in simulations.

17.3 Results and Discussion

Three CFD simulations were performed according to the previous description. Experimental data from the opaque wall collector tube used for comparison with simulations are listed in Table 17.1.

17.3.1 Pressure Drop

The three systems were simulated towards pressure drop as a function of time, from the fixed to fluidized bed regime. As expected, the pressure drop average values increase as particles fill the tube and it finally achieves a stable average value when the SiC particles reach the tube exit. It reaches 32,000 Pa for Run 1, 25,000 Pa for Run 2, and 23,000 Pa for Run 3, the difference being due to the average solid fraction in the tube in a pseudo-stationary condition. In heated systems (Runs 2 and 3), the air expansion due to temperature increase reduces the average solid fractions, and total pressure drops are lower than in cold systems. The stationary values of pressure drop obtained from simulations are in good agreement with experimental ones of respectively 31,931 Pa (cold experimental run) and 25,381 Pa (hot experimental runs). As the tube is exhausted to the atmosphere, the net power required is the sum of hydrostatic, acceleration and friction losses, as defined by Zhang et al. (2016).

17.3.2 Wall Heat Exchange

Figure 17.3 shows the time-dependent net power flux densities transferred to the suspension. Initially, when there is only air in the heated zone, the power flux is very low. When the solids reach this region, the thermal flux increases until it finally achieves pseudo-stability. In terms of power, this state is achieved in considerably less time (about half the time) at the high solid mass flow rate in comparison with the low solid mass flow rate.

Figure 17.4 shows mean particle temperature in planes located at the entrance and exit of the heated zone as a function of time for the simulation with high (Run 2) and low (Run 3) SiC mass flow rate, respectively, at a wall temperature of 509.4 and 442.5 K, respectively.

Initially, in both simulations, the average temperature in the plane located at the entrance of the heated zone remains constant and is equal to the value imposed as

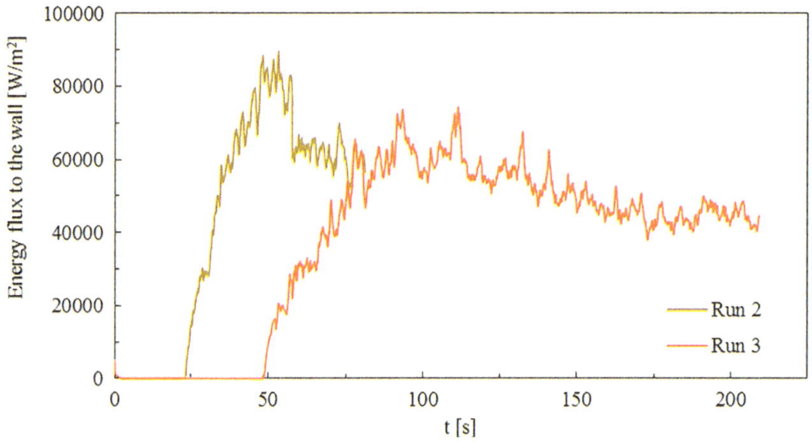

Fig. 17.3 Instantaneous average net power flux on the hot wall, for Run 2 and Run 3

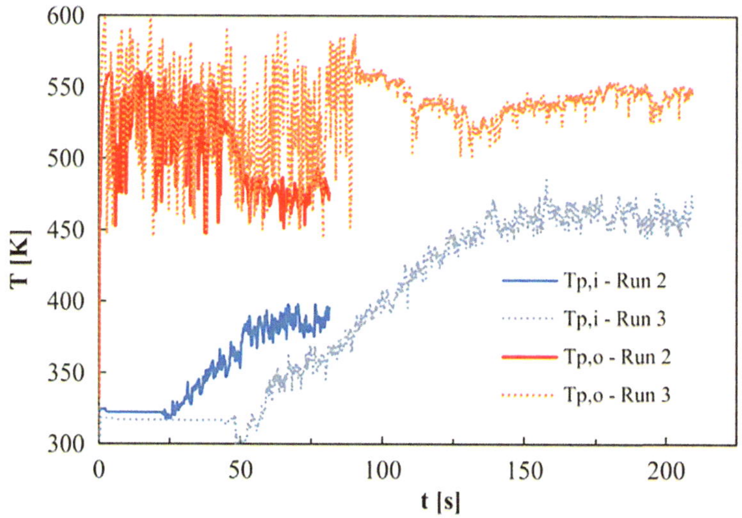

Fig. 17.4 Evolution of particle average temperature, T_p, in planes located at the entrance ($T_{p,i}$) and exit ($T_{p,o}$) of the heat transfer area of the system

Table 17.3 Results of CFD simulations in pseudo-stationary state (mean values) and comparison between experimental and simulation results

	$\varepsilon_{p,exp}$	$\varepsilon_{p,CFD}$	$T_{p,i,CFD}$ (K)	$T_{p,i,exp}$ (K)	$T_{p,o,CFD}$ (K)	$T_{p,o,exp}$ (K)
Run 1	0.39	0.38	–	–	–	–
Run 2	0.31	0.33	385.50	381.4–389.6	492.67	488.5–495.8
Run 3	0.31	0.32	460.82	455.0–465.9	545.47	542.1–545.8

boundary condition. This behavior is maintained for 25 s in Run 2, and for 50 s in Run 3. Then, the particle temperature increases steadily as a result of the recirculation of solid particles from the top towards the bottom of the heated zone. Finally, temperature pseudo-stability is achieved, starting from 80 s for Run 2 and 150 s for Run 3.

To compare experimental and CFD results, Table 17.3 summarizes the pseudo-stationary average values of the solid volume fraction and the solid temperature at the inlet and outlet planes of the heated region. Absolute errors for $T_{p,i}$ and $T_{p,o}$ and relative errors for ε_p refer to experimental values from Table 17.1.

$$\varepsilon_p = \frac{\Delta P_{tube}}{(\rho_p - \rho_g)Lg} \quad (17.1)$$

Clearly, the model fits both the solid volume fraction for all conditions analyzed and the particle temperatures (inlet/exit) in both mass flow rate conditions (Run 2 and Run 3).

17.3.3 *Slug Formation in the Tube*

The simulations show the presence of wall slugs in the tube for all simulated operating conditions. Figure 17.5 shows the iso-surface of the air volume fraction (ε_g) of 0.8, value assigned to the bubble/slug-dense phase interface. This value was adopted following the work of Lu et al. (2015).

Figure 17.5 demonstrates that the slugs are generated at a certain height above the entrance of the tube (indicated by the blue circle) and they initially maintain a relatively small size in comparison to that acquired in the heat transfer zone, where they occupy most of the tube wall and tube cross sectional area. It is however important to assess the influence of the temperature on slugs' size and thickness. Regarding their size, in the hot zone the slugs occupy extensive regions of the wall tube surface, covering approximately 50%. With respect to the slugs' thickness, we can observe average values of 2 mm in the insulated region, while in the heated area thicknesses

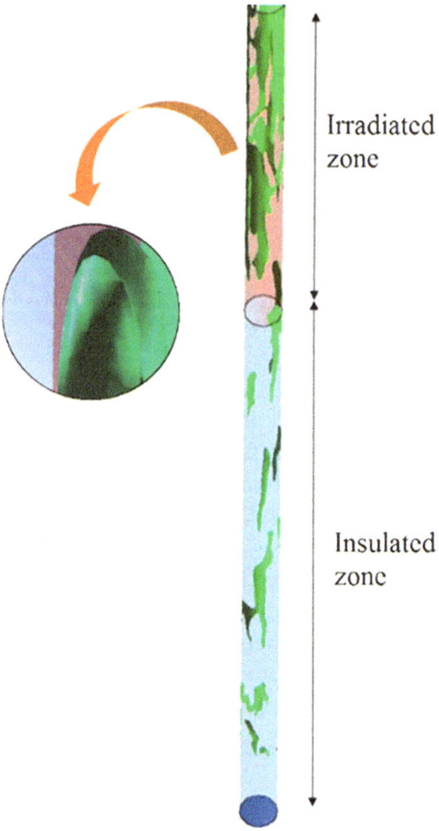

Fig. 17.5 Presence of slugs in the area adjacent to the tube wall for Run 2. In blue, the plane of admission of the emulsion to the domain; in red, the plane of exit of heat transfer zone

between 4 and 6 mm are observed. Slug velocities could be determined from successive images. They will be reported upon and compared with Kong et al. (2017) values in a follow up paper.

17.3.4 Analysis of Radial Variables

Four selected planes in the radial direction (see Fig. 17.6) were discretized in order to evaluate both the mean velocity profile of the granular solid and of the air, and the average volume fraction of the granulated material as a function of the tube height. The average temperature profiles of SiC particles are also plotted for hot experiments. The cross sections (planes) indicated in Fig. 17.6 correspond to the positions $z = 0.40$ m (Level 1), 0.80 m (Level 2), 1.05 m (Level 3) and 1.20 m (Level 4).

In the discretized planes, the thickness of each ring was defined as a function of the radial position. Near the wall, where particle conduction heat transfer prevails, a 0.5 mm ring was used and this value was increased towards the tube center.

Figure 17.7 show the variables ε_p, $v_{p,z}$ and $v_{g,z}$ obtained from the cold simulation. According to Fig. 17.7b, two clearly distinguishable zones exist: (1) a zone where particles flow upward in the center of the tube and (2) an annular region between the wall and the rising zone, in which the particles flow down with a particle volume fraction greater than the average value (Fig. 17.7a). A very thin region at the wall is predicted by the simulation with particles flowing upward. Figure 17.7c shows a similar behavior for the axial gas velocity values. Therefore, we can conclude that in the radial domain the particles move along with the gas.

The results for Run 2 are also shown in Fig. 17.7 to be compared with the cold case. Figure 17.7b, c show that the slugs' thickness increases with axial temperature. The velocity of both phases is higher in the region adjacent to the hot wall than in the lower cold zone of the riser. This is due to air density changes with temperature increase in the axial direction.

At level 1 (Fig. 17.6), located in the insulated area of the tube, slugs are also present in the vicinity of the wall. Similar results are obtained for Run 3 although the mean axial velocities in the same region are negative, so, apparently, there are no slugs present in this region. The difference is due to the frequency at which the slugs appear for the two analyzed SiC mass flow rates. For low SiC mass flow rate, the slugging frequency is lower than observed at higher mass flow rate, which results in predominantly negative axial velocity.

Figure 17.8a (Run 3) and b (Run 2) show the instantaneous z velocity component for both conditions. For low SiC mass flow rate, the average velocity is predominantly negative throughout the interval analyzed. Positive axial velocity peaks appear when a slug flows in the analyzed region. Contrarily, for a high mass flow rate, the axial velocity is mainly positive because of the predominant slug occurrence.

Figure 17.9 shows particle velocity vectors and solid volume fraction in the ring adjacent to the wall (Level 1) under pseudo-stationary condition. The areas with low volume concentration of solids correspond to regions where slugs rise and where the

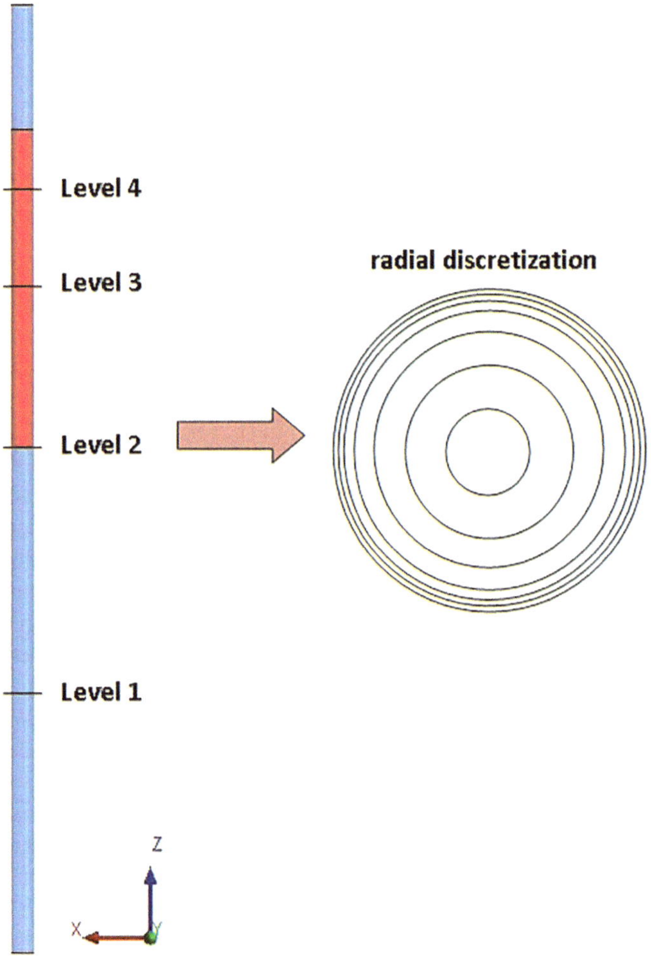

Fig. 17.6 Geometry of the simulated tube (left). In red, the region of the tube that receives concentrated solar energy. In the center, the tube in perspective. To the right, the discretization of each plane

magnitude of particle velocity reaches maximum relative values. The zones of high concentration of solids present low magnitude velocities and downward direction. The surface covered with slugs observed in Run 2 is greater than that observed in Run 3.

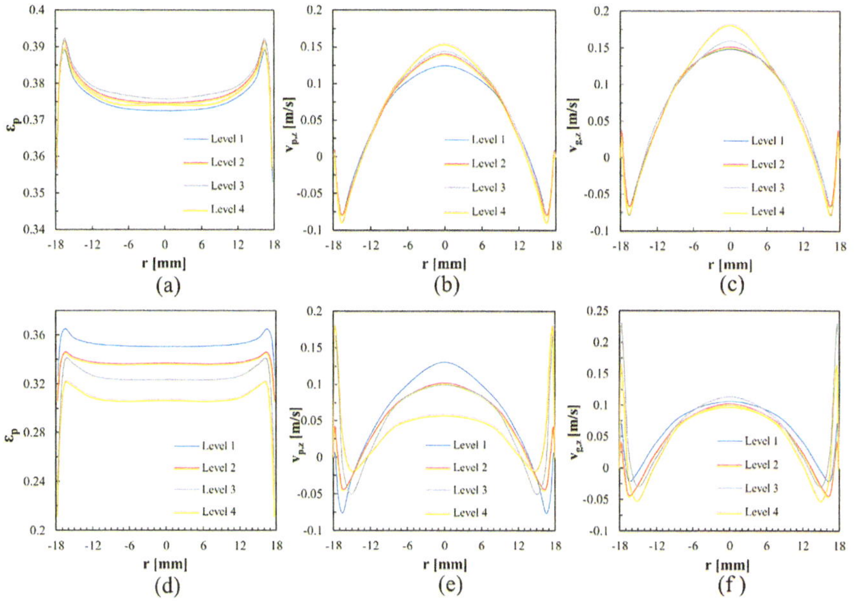

Fig. 17.7 Axial ε_p radial profile for Run 1 (**a**) and Run 2 (**d**); particle velocity ($v_{p,z}$) (**b**, **e**); air velocity (**c**, **f**)

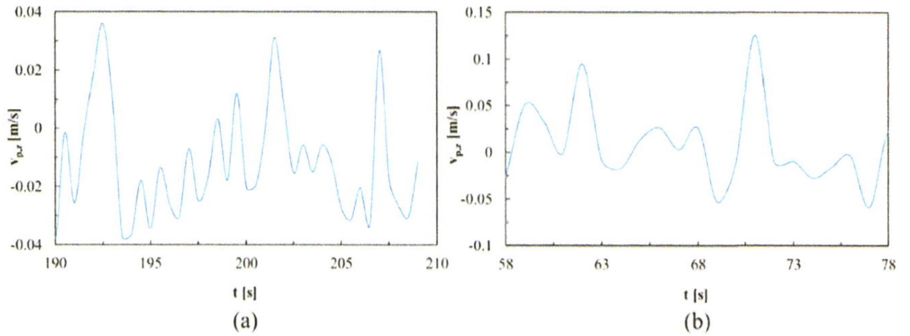

Fig. 17.8 Instant z-velocity component in the zone adjacent to the wall at Level 1. **a** Run 3; **b** Run 2

17.3.5 Analysis of Axial Variables

Figure 17.10a shows mean values of solid volume fraction at different heights. Between the bottom and 0.2 m, there is rapid increase of the solid volume fraction for all cases. After 0.6 m, the volume fraction is rather constant for run 1 (cold case) and it decreases for run 2 and 3 due to the expansion of fluidization air (temperature increase with height in the heating zone).

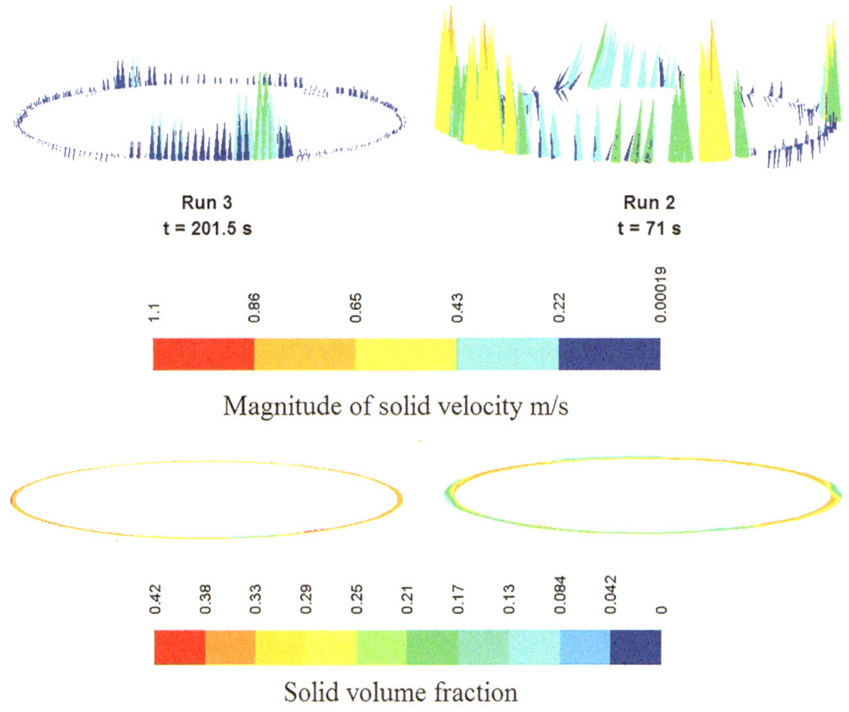

Fig. 17.9 Velocity vectors and contours of the granular solid phase corresponding to level 1 (insulated zone) in region adjacent to the wall

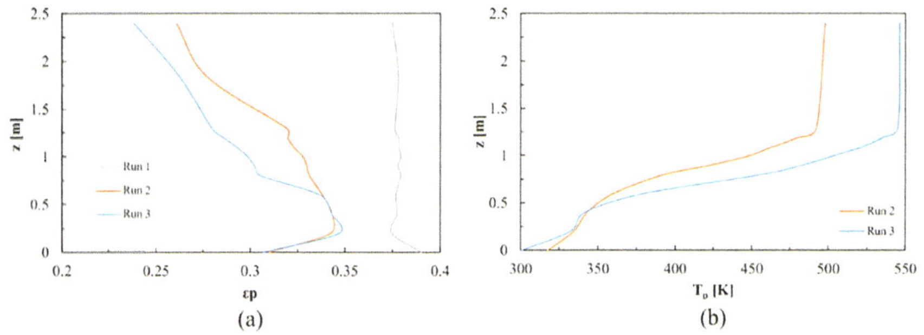

Fig. 17.10 **a** Volume fraction occupied by SiC particles for different positions of the axial coordinate; **b** mean temperature of SiC particles for different positions of *z* coordinate

Figure 17.10b shows the mean particle temperature as a function of the axial position. The reflux of particles in the tube results in a temperature increase in almost the entire lower insulated region.

17.3.6 Influence of Slug Formation on the Wall-to-Suspension Heat Transfer

In order to analyze the dynamics of wall slugs and their influence on the heat flux transferred along the wall, three 0.5 m long volumes each along the tube cross sectional area located at different heights were selected. The first one is situated in the lower part of the tube's heated zone, the second one is in the center, and the third one is close to the outlet region of the heated zone, hence all between 0.8 and 1.3 m of the tube. In these volumes, we evaluated the wall heat flux and the tube surface fraction covered by slugs.

Figure 17.11 show the average energy flux and the percentage of the tube wall covered by slugs for Run 2 and Run 3 respectively. With respect to heat flux, the value of this magnitude is estimated by the software ANSYS Fluent 17, based on the temperature profile at the wall for each region of the tube. According to the constitutive equation for effective conductive heat flux, the value of the effective thermal conductivity is in straight relation to the resulting heat flux.

It can be observed that in both simulations the heat flux density decreases with height by a factor of about two along 0.5 m. This decrease of net power flux transferred to the suspension is due to three effects, all related to the presence of wall slugs: the decrease of the driving force for unsteady state heat transfer due to the less frequent particle removal at the inner tube wall, the throughflow of gas from slug to slug thus convectively decreasing the prevailing temperature difference between slugs and tube wall, and the slugs' lower thermal conductivity and lower heat capacity in comparison with the dense phase (Baeyens and Geldart 1980; Mazza et al. 1997).

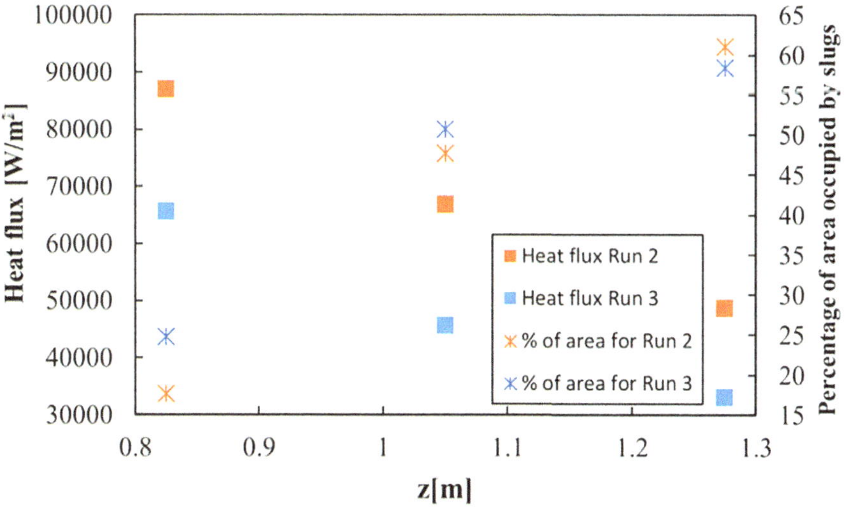

Fig. 17.11 Heat flux and internal wall fraction covered by slugs as a function of tube height for Run 2 and Run 3

For this reason, the slugs act as a resistance for heat transfer causing low values of mean heat flux exchanged between the wall and the fluid bed.

17.4 Conclusions

This study investigates the fluid-dynamics of dense upflow bubbling fluidized systems of Geldart A particles in small diameter beds based on CFD simulation technique. The model validation was carried out considering experimental results reported in previous contributions with reference to a prototype operated at the PROMES-CNRS laboratory.

The resulting inlet and outlet temperatures, $T_{p,i}$ and $T_{p,o}$, provided a way to validate the fluid-dynamic behavior of the fluidized dense particulate medium.

The main findings of this work can be summarized as follows:

- The horizontal zone of each fluidization curve is used to determine the average solid fraction in the tube in a pseudo-stationary condition.
- The stationary values of pressure drop obtained from simulations are in good agreement with experimental ones, as can be verified through ε_p values presented in Table 17.3.
- The model also reproduces well the particle temperature at the entrance and the exit of the heated zone, in both simulations including heat transfer.
- The simulations show the presence of wall slugs in the tube for all simulated operating conditions.
- It is observed that the slugs are generated at a certain height above the entrance of the tube and, initially, they maintain a relatively small size in comparison to that acquired in the heat transfer zone, where they get to occupy most of the tube surface.
- The influence of temperature on slugs' size and thickness is considerable. In the hot zone the slugs occupy extensive regions of the wall tube surface, covering approximately 50% of it. With respect to the slugs' thickness, we can observe average values of 2 mm in the insulated region, while in the heated area thicknesses between 4 and 6 mm are observed.
- Solids movement follows a logic pattern induced by the bubbles. It consists of a common bubbling fluidized bed particle upward/downward behavior as previously reported and producing upward movement of both, solid and gas phases in the center of the column. In the bubble-lean zone of the bed, both phases flow downwards.
- It is demonstrated from simulations validated by experimental temperature measurements that the slugs act as a resistance for heat transfer causing low values of mean heat flux exchanged between the wall and the fluid bed.

The results emphasize the effect of the hot wall on slug's formation and size. It can also be observed that the slug thickness increases in this area. These findings are particularly relevant for heat transfer analysis because slugs' formation strongly hampers heat transfer from the tube wall to the bed.

Nomenclature

C_D	Drag coefficient, dimensionless
C_p	Heat capacity, J kg^{-1} K^{-1}
C_1, C_2	Characteristic coefficients in Syamlal and O'Brien model
d_p	Mean Sauter diameter of SiC particles, m
$d_{p,eff}$	Effective diameter of SiC particles, m
e_{pp}	Restitution coefficient, dimensionless
$g_{0,p}$	Radial distribution function of solid, dimensionless
k_b	Effective thermal conductivity of the emulsion, W m^{-1} K^{-1}
$k_{b,g}$	Effective thermal conductivity of the gas phase, W m^{-1} K^{-1}
$k_{b,p}$	Effective thermal conductivity of the solid phase, W m^{-1} K^{-1}
$k_{p,o}$	Thermal conductivity of SiC, W m^{-1} K^{-1}
$k_{g,o}$	Thermal conductivity of air (temperature dependent), W m^{-1} K^{-1}
$K_{g,p}$	Momentum exchange coefficient, kg m^{-3} s^{-1}
$\dot{m}_{SiC}$	Mass flow rate of SiC, kg h^{-1}
$Nu_{g,p}$	Nusselt number for gas–solid systems
P	Pressure, Pa
Pr	Prandtl number
P_p	Solid phase pressure, Pa
$\vec{q}_g$	Energy exchanged between gas phase and the tube wall, W m^{-2}
$Q_{g,p}$	Energy exchanged between phases, W m^{-2}
$\dot{Q}$	Net power flux transferred to the suspension, W m^{-2}
Re_p	Relative Reynolds number
$\vec{R}_{g,p}$	Force vector of the drag between phases per volume unit, N m^{-3}
$T_{p,i}$	Particles temperature at the entry of the heat transfer zone, K
$T_{p,o}$	Particles temperature at the exit (upper part of the tube), K
$T_{W,Z}$	External temperature function at the tube wall, K
$T_{p,cold}$	Particles temperature at the entry of the riser, K
$\vec{u}_g$	Interstitial velocity vector of the gas, m s^{-1}
$\vec{u}_p$	Interstitial velocity vector of the particles, m s^{-1}
$u_{t,p}$	Terminal velocity of SiC particles, m s^{-1}
U_{mb}	Superficial minimum bubbling velocity, m s^{-1}
U_{mf}	Superficial minimum fluidization velocity, m s^{-1}
V_q	Volume occupied by a q generic phase, m^3
V	Cell volume, m^3
z	Axial coordinate, m

Greek symbols

ε_{mf}, ε_{mf}	Voidage of the bed at U_{mf} and U_{mb}, respectively, dimensionless
ε_p	Volumetric fraction occupied by SiC particles, dimensionless
$\varepsilon_{p,max}$	Maximum acceptable solid fraction, dimensionless
ε_g	Volumetric fraction occupied by the gas, dimensionless
μ_g	Gas viscosity, Pa s
μ_p	Granular phase viscosity, Pa s
ρ_B	Bulk density of the fluidized bed, kg m^{-3}
ρ_p	Absolute particle density, kg m^{-3}
ρ_g	Density of air, kg m^{-3}
$\bar{\bar{\tau}}_g$	Shear stress tensor of the gas phase, Pa
ϕ_p	Sphericity of SiC particles, dimensionless
$\nu_{t,p}$	Ratio of the terminal settling velocity of a multi-particle system to that of an insulated single particle, dimensionless

Appendix

See Tables 17.4 and 17.5.

Table 17.4 Governing and constitutive equations of the Euler-Euler model

Definition	
$V_i = \int_V \varepsilon_i dV \text{with} \varepsilon_g + \varepsilon_p = 1$	(17.2)
Mass conservation	
$\frac{\partial}{\partial t}(\varepsilon_g \rho_g) + \nabla.(\varepsilon_g \rho_g \overrightarrow{u}_g) = 0$	(17.3a)
$\frac{\partial}{\partial t}(\varepsilon_p \rho_p) + \nabla.(\varepsilon_p \rho_p \overrightarrow{u}_p) = 0$	(17.3b)
Momentum conservation	
$\frac{\partial}{\partial t}(\varepsilon_g \rho_g \vec{u}_g) + \nabla \cdot (\varepsilon_g \rho_g \vec{u}_g \vec{u}_g) = -\varepsilon_g \nabla P + \nabla \cdot \bar{\bar{\tau}}_g + \varepsilon_g \rho_g \vec{g} + \vec{R}_{g,p}$	(17.4a)
$\frac{\partial}{\partial t}(\varepsilon_p \rho_p \vec{u}_p) + \nabla \cdot (\varepsilon_p \rho_p \vec{u}_p \vec{u}_p) = -\varepsilon_p \nabla P + \nabla P_p + \nabla \cdot \bar{\bar{\tau}}_p + \varepsilon_p \rho_p \vec{g} + \vec{R}_{p,g}$	(17.4b)
With $\vec{R}_{g,p} = \vec{R}_{g,p}$ (17.4c) $\vec{R}_{g,p} = K_{g,p}(\vec{u}_g - \vec{u}_p)$	(17.4d)
$P_p = \varepsilon_p \rho_s \Theta_p + 2\rho_p(1 + e_{pp})\varepsilon_p^2 g_{(0,p)} \Theta_p$	(17.4e)
$\bar{\bar{\tau}}_i = \varepsilon_i \mu_i \left(\nabla \vec{u}_i - \nabla \vec{u}_i^T\right) + \varepsilon_i(\lambda_i - \frac{2}{3}\vec{u}_i)\nabla \cdot \vec{u}_i \bar{\bar{I}}$	(17.4f)
Energy conservation	
$\frac{\partial}{\partial t}(\varepsilon_g \rho_g h_g) + \nabla \cdot (\varepsilon_g \rho_g \vec{u}_g h_g) = \varepsilon_g \frac{\partial P_g}{\partial t} + \bar{\bar{\tau}}_g : \nabla \vec{u}_g + \nabla \cdot \vec{q}_g + Q_{g,p}$	(17.5a)
$\frac{\partial}{\partial t}(\varepsilon_p \rho_p h_p) + \nabla \cdot (\varepsilon_p \rho_p \vec{u}_p h_p) = \varepsilon_p \frac{\partial P_p}{\partial t} + \bar{\bar{\tau}}_p : \nabla \vec{u}_p + \nabla \cdot \vec{q}_p + Q_{p,g}$	(17.5b)
with $Q_{g,p} = Q_{p,g}$(17.5c); $h_i = \int_{T_{ref}}^{T} c_{p,i} dT_i$(17.5d); $\nabla.\overrightarrow{q}_i = \nabla.(-k_i \nabla T_i)$	(17.5e)
Solid shear viscosity $\mu_p = \mu_{p,col} + \mu_{p,kin} + \mu_{p,fr}$	(17.6)
Collisional viscosity ($\mu_{p,col}$) (Syamlal and O'Brien 1987) $\mu_{(p,col)} = \frac{4}{5}\varepsilon_p^2 \rho_p d_p g_{(0,p)}(1 + e_{pp})\left(\frac{\Theta_p}{\pi}\right)^{1/2}$	(17.7)
Kinetic viscosity $\mu_{p,kin} = \frac{\varepsilon_p d_p \rho_p \sqrt{\Theta_p \pi}}{6(3-e_{pp})}\left[1 + \frac{2}{5}(1 + e_{pp})(3e_{pp} - 1)\varepsilon_p g_{0,p}\right]$	(17.8)
Frictional viscosity ($\mu_{p,fr}$): not necessary to be taken into account since ε_p is less than the maximum allowed ($\varepsilon_{p,max}$) of 0.63 in every case (Gunn 1978)	
Bulk viscosity within the bed (Lun et al. 1984) $\lambda_p = \frac{4}{3}\varepsilon_p \rho_p d_p g_{0,pp}(1 + e_{pp})\left(\frac{\Theta_p}{\pi}\right)^{1/2}$	(17.9)
Radial distribution function (Ogawa et al. 1980) $g_{0,p} = \left[1 - \left(\frac{\varepsilon_p}{\varepsilon_{p,max}}\right)^{1/3}\right]^{-1}$	(17.10)
Solid pressure $P_s = 2\rho_s(1 + e_{pp})\varepsilon_p^2 g_{0,p}\Theta_p$	(17.11)

Table 17.5 Exchange equations between phases and Zehner and Schlünder model

Momentum interphase exchange coefficient	
$K_{g,p} = \frac{3}{4}\frac{\varepsilon_p\varepsilon_g\rho_g}{u_{t,p}^2 d_{p,eff}} C_D\left(\frac{Re_p}{u_{t,p}}\right)\left\|\vec{u}_p - \vec{u}_g\right\|$	(17.12)
$d_{p,eff} = \phi_p{}^n d_p n = 2.53$	(17.13)
$C_D = \left(0.63 + \frac{4.8}{\sqrt{Re_p/\nu_{t,p}}}\right)^2$ (DallaValle 1948)	(17.14)
$\nu_{t,p} = 0.5\left(A - 0.06Re_p + \sqrt{(0.06Re_p)^2 + 0.12Re_p(2B - A) + A^2}\right)$ (Garside and Al-Dibouni 1977)	(17.15)
$A = \varepsilon_g^{4.14}$ and $B = C_1\varepsilon_g^{1.28}$ for $\varepsilon_g \leq 0.85$ or $B = \varepsilon_g^{C_2}$ for $\varepsilon_g > 0.85$ with $C_1 = 0.8$ and $C_2 = 2.65$	(17.16)
Convective heat transfer coefficient	
$h_{g,p} = \frac{6k_{g,o}\varepsilon_p\varepsilon_g Nu_{g,p}}{d_p^2}$	(17.17)
$Nu_{gp} = \left(7 - 10\varepsilon_g + 5\varepsilon_g^2\right)\left(1 + 0.7Re_p^{0.2}Pr^{\frac{1}{3}}\right) + \left(1.33 - 2.4\varepsilon_g + 1.2\varepsilon_g^2\right)Re_p^{0.7}Pr^{1/3}$ (Gunn 1978)	(17.18)
Effective dense suspension conductivity (Kang et al. 2019): $k_b = k_{b,g} + k_{b,p}$	(17.19)
Effective gas conductivity: $k_{b,g} = \left(1 - \sqrt{1 - \varepsilon_g}\right)k_{g,o}$	(17.20)
Effective conductivity of granular medium (Kang et al. 2019): $k_{b,p} = \sqrt{1 - \varepsilon_g}[\omega A + (1 - \omega)\Gamma]k_{g,o}$ with $\omega = 7.26x10^{-3}$	(17.21)
$\Gamma = \frac{2}{\left(1-\frac{B}{A}\right)}\left\{\frac{(A-1)}{\left(1-\frac{B}{A}\right)^2}\frac{B}{A}\ln\left(\frac{A}{B}\right) - \frac{(B-1)}{\left(1-\frac{B}{A}\right)} - \frac{1}{2}(B + 1)\right\}$	(17.22)
$B = 1,25\left(\frac{(1-\varepsilon_g)}{\varepsilon_g}\right)^{10/9}$ and $A = \frac{k_{p,o}}{k_{g,o}}$	(17.23)

References

Armstrong LM, Gu S, Luo KH (2010) Study of wall-to-bed heat transfer in a bubbling fluidised bed using the kinetic theory of granular flow. Int J Heat Mass Transf 53:4949–4959

Baeyens J, Geldart D (1974) An investigation into slugging fluidized beds. Chem Eng Sci 29:255–265

Baeyens J, Geldart D (1980) Modelling approach to the effect of equipment scale on fluidised bed heat transfer data. J Powder Bulk Solids Technol 4:1

DallaValle JM (1948) Micromeritics: the technology of fine particles

Deng Y, Sabatier F, Dewil R, Flamant G, Le Gal A, Gueguen R, Baeyens J, Li S, Ansart R (2021) Dense upflow fluidized bed (DUFB) solar receivers of high aspect ratio: different fluidization modes through inserting bubble rupture promoters. Chem Eng J 418:129376

Garside J, Al-Dibouni MR (1977) Velocity-voidage relationships for fluidization and sedimentation in solid-liquid systems. Ind Eng Chem Process Des Dev 16:206–214

Gidaspow D, Bezburuah R, Ding J (1992) Hydrodynamics of circulating fluidized beds: Kinetic theory approach. In: Conference: 7th international conference on fluidization, Gold Coast (Australia). United States

Gunn DJ (1978) Transfer of heat or mass to particles in fixed and fluidised beds. Int J Heat Mass Transf. https://doi.org/10.1016/0017-9310(78)90080-7

Kang Q, Flamant G, Dewil R, Baeyens J, Zhang HL, Deng YM (2019) Particles in a circulation loop for solar energy capture and storage. Particuology 43:149–156

Kong W, Tan T, Baeyens J, Flamant G, Zhang H (2017) Bubbling and slugging of Geldart group a powders in small diameter columns. Ind Eng Chem Res 56:4136–4144

Kong W, Baeyens J, Flamant G, Tan T, Zhang H (2019) Solids flow in a "particle-in-tube" concentrated solar heat absorber. Ind Eng Chem Res 58:4598–4608

Kuipers JAM, Prins W, Van Swaaij WPM (1992) Numerical calculation of wall-to-bed heat-transfer coefficients in gas-fluidized beds. AIChE J 38:1079–1091

Lu Y, Huang J, Zheng P (2015) A CFD–DEM study of bubble dynamics in fluidized bed using flood fill method. Chem Eng J 274:123–131

Lun CKK, Savage SB, Jeffrey DJ, Chepurniy N (1984) Kinetic theories for granular flow: inelastic particles in Couette flow and slightly inelastic particles in a general flowfield. J Fluid Mech 140:223–256

Mazza GD, Mariani NJ, Barreto GF (1997) Evaluation of overall heat transfer rates between bubbling fluidized beds and immersed surfaces. Chem Eng Commun 162:125–149

Ogawa S, Umemura A, Oshima N (1980) On the equations of fully fluidized granular materials. Zeitschrift für Angew Math und Phys ZAMP 31:483–493

Syamlal M, O'Brien TJ (1987) The derivation of a drag coefficient formula from velocity-voidage correlations. Tech Note, US Dep energy, Off Foss Energy, NETL, Morgantown, WV

Taghipour F, Ellis N, Wong C (2005) Experimental and computational study of gas–solid fluidized bed hydrodynamics. Chem Eng Sci 60:6857–6867

Zhang H, Benoit H, Gauthier D, Degrève J, Baeyens J, López IP, Hemati M, Flamant G (2016) Particle circulation loops in solar energy capture and storage: gas–solid flow and heat transfer considerations. Appl Energy 161:206–224

Zhang H, Benoit H, Perez-Lopèz I, Flamant G, Tan T, Baeyens J (2017) High-efficiency solar power towers using particle suspensions as heat carrier in the receiver and in the thermal energy storage. Renew Energy 111:438–446

Zhang H, Baeyens J, Dewil R, Li S, Deng Y, Reyes-Urrutia A, Venier Cesar M, Mazza G (2021) Solar biomass pyrolysis with latent heat ballast. In: 2021 5th international conference on power and energy engineering (ICPEE). IEEE, pp 85–90

The manufacturer's authorised representative in the EU is Springer Nature Customer Service Centre GmbH, Europaplatz 3, 69115 Heidelberg, Germany. If you have any concerns regarding our products, please contact ProductSafety@springernature.com

Printed and bound by CPI Group (UK) Ltd, Croydon, CR0 4YY
01/10/2026
02225868-0003